ENGINEERING ASPECTS OF THERMAL POLLUTION

Engineering

Aspects of

THERMAL
POLLUTION

PROCEEDINGS OF THE NATIONAL SYMPOSIUM ON THERMAL POLLUTION
SPONSORED BY FEDERAL WATER POLLUTION CONTROL ADMINISTRATION
AND VANDERBILT UNIVERSITY
NASHVILLE, TENNESSEE, AUGUST 14-16, 1968

Edited by

Frank L. Parker AND
Peter A. Krenkel

VANDERBILT UNIVERSITY PRESS • 1969

Printed in the United States of America
Standard Book Number 8265–1145–7
Library of Congress Catalogue Card Number 79-92266

This symposium and the National Symposium on Biological Aspects of Thermal Pollution which preceded it on June 3–5, 1968, in Portland, Oregon, were supported in part by funds from the Federal Water Pollution Control Administration, U.S. Department of the Interior.

PREFACE

THE problem of thermal pollution is expounded upon daily. There are, perhaps, two main reasons for this heightened attention though, in fact, very little damage due to thermal pollution has been observed, as yet. First, the problem is growing at a geometric rate so that by the year 2000 the equivalent of the total flow of surface water in the United States will be needed for cooling purposes if once-through cooling is required. Second, because very little damage has been done so far, there is the opportunity offered the electrical utility industry, potentially the major offender, to avoid what very few industries have done so far and that is to solve the pollution problem before it becomes onerous.

As in all solutions of real-life problems, the benefits to be achieved by abating pollution must be balanced against the costs to arrive at a socially acceptable mixture of abatement and pollution. To arrive at the optimum solution would require writing an objective function which in turn demands complete knowledge of the effects, both deleterious and beneficial, of thermal pollution and the costs of avoiding them.

This complete knowledge is not available and even what is known is scattered widely in the biological, engineering, and economic literature. Therefore, the Federal Water Pollution Control Administration granted Vanderbilt University funds to cosponsor two symposia on thermal pollution to bring together the available knowledge in one convenient source. Proceedings of the first symposium, published under the title *Biological Aspects of Thermal Pollution,* include the formal papers and discussions presented at Portland, Oregon, on June 3–5, 1968. Proceedings of the second symposium, held at Vanderbilt University in Nashville, Tennessee, on August 14–16, 1968, dealt with engineering aspects of thermal pollution and appear in this volume.

The format of the symposia, with distinguished invited lecturers and respondees, gave ample time for floor and informal discussion. The floor discussion, as edited, is included in each volume. The major thrust of the symposia was to bring together well-informed people from the universities, industry, regulatory agencies, and engineering to discuss the problem, elicit comments, and formulate a program of research. I believe that the objectives of the symposia have been achieved.

Co-chairmen of the Nashville symposium were James R. Boydston of Pacific Northwest Water Laboratory, Federal Water Pollution Control Administration, and Peter A. Krenkel, professor and chairman of the Department of Environmental and Water Resources Engineering of Vanderbilt University.

Thanks are due to personnel of the FWPCA: William A. Cawley, J. Frances Allen, Bruce Tichenor, and Frank Rainwater; and of Vanderbilt University: Edward L. Thackston, Priscilla Smith, Elizabeth Fletcher, and Maryann Moore, for their help in producing the symposia and this volume. Major thanks are also due to the writers of these papers and to those who contributed to the floor discussions.

Vanderbilt University   FRANK L. PARKER
May 1969   PETER A. KRENKEL

THE AUTHORS

PETER ACKERS
: Senior Principal Scientific Officer, Hydraulic Research Station, Ministry of Technology, Wallingford, Berkshire, England

DANIEL W. BATES
: U.S. Bureau of Fisheries, Portland, Oregon

JACOB I. BREGMAN
: U.S. Deputy Assistant Secretary for Water Pollution Control, U.S. Department of the Interior, Washington, D.C.

NORMAN BROOKS
: Professor of Civil Engineering, California Institute of Technology, Pasadena, California

ROBERT S. BURD
: Director, Water-Quality Standards, Office of Program Planning and Development, Federal Water Pollution Control Administration, Washington, D.C.

WILLIAM A. CAWLEY
: Acting Director, Pollution Control Technology Branch, Division of Research, Federal Water Pollution Control Administration, U.S. Department of the Interior, Washington, D.C.

MILO A. CHURCHILL
: Chief, Water-Quality Branch, Tennessee Valley Authority, Chattanooga, Tennessee

EDWARD E. DRIVER
: Research Engineer, Engineering Laboratory, Tennessee Valley Authority, Norris, Tennessee

JOHN E. EDINGER
: Assistant Professor of Sanitary and Water Resources Engineering, Vanderbilt University, Nashville, Tennessee

REX A. ELDER
: Director, Engineering Laboratory, Tennessee Valley Authority, Norris, Tennessee

CARLOS FETTEROLF
> Chief, Water-Quality Appraisal Unit, Michigan Water Resources Commission, Lansing, Michigan

GORDON L. FORD
> Senior Mechanical Design Engineer, Tennessee Valley Authority, Knoxville, Tennessee

ANDRE GOUBET
> Director, Paris District, Electricite de France, Paris, France

G. EARL HARBECK, JR.
> Research Hydrologist, U.S. Geological Survey, Denver, Colorado

DONALD R. F. HARLEMAN
> Professor of Hydraulic Engineering, Massachusetts Institute of Technology, Cambridge, Massachusetts

H. A. HAWKES
> Lecturer, Department of Biological Sciences, University of Aston at Birmingham, England

PETER A. KRENKEL
> Chairman, Department of Environmental and Water Resources Engineering, Vanderbilt University, Nashville, Tennessee

GEORGE O. G. LÖF
> Professor of Civil Engineering, Colorado State University, Fort Collins, Colorado

JOE G. MOORE
> Commissioner, Federal Water Pollution Control Administration, U.S. Department of the Interior

LORING F. OEMING
> Executive Secretary, Michigan Water Resources Commission, Lansing, Michigan

FRANK L. PARKER
> Professor of Environmental and Water Resources Engineering, Vanderbilt University, Nashville, Tennessee

W. R. SHADE
> Chief Mechanical Engineer, Gilbert and Associates, Reading, Pennsylvania

ALEX F. SMITH III
> Assistant to Chief Engineer, Gilbert and Associates, Reading, Pennsylvania

BRUCE A. TICHENOR
> Research Sanitary Engineer, National Thermal Pollution Research Program, Federal Water Pollution Control Administration, Pacific Northwest Water Laboratory, Corvallis, Oregon

JOHN C. WARD

Associate Professor of Civil Engineering, Colorado State University, Fort Collins, Colorado

CHARLES WASELKOW
Manager of Production, Public Service Company of Colorado, Denver, Colorado

EUGENE B. WELCH
Supervisor, Biological Section, Tennessee Valley Authority, Chatta-nooga, Tennessee

CHAIRMEN OF THE MEETINGS

LEON W. WEINBERGER
> Assistant Commissioner, Research and Development, Federal Water Pollution Control Administration, Washington, D.C.

FRANCIS E. GARTRELL
> Assistant Director of Health, Tennessee Valley Authority, Chattanooga, Tennessee

FRANK H. RAINWATER
> Chief, National Thermal Pollution Research, Federal Water Pollution Control Administration, Corvallis, Oregon

THE SYMPOSIUM COMMITTEE
National Symposium on Thermal Pollution

JAMES L. AGEE
    Director, Pacific Northwest Laboratory, Federal Water Pollution Control Administration, Corvallis, Oregon

JAMES R. BOYDSTON
    Director, Treatment and Control Research, Pacific Northwest Water Laboratory, Federal Water Pollution Control Administration, Corvallis, Oregon

WILLIAM A. CAWLEY
    Acting Director, Pollution Control, Technology Branch, Division of Research, Federal Water Pollution Control Administration, U.S. Department of the Interior, Washington, D.C.

PETER A. KRENKEL
    Chairman, Department of Environmental and Water Resources Engineering, Vanderbilt University, Nashville, Tennessee

FRANK L. PARKER
    Professor of Environmental and Water Resources Engineering, Vanderbilt University, Nashville, Tennessee

CO-CHAIRMEN OF THE COMMITTEE:
    JAMES R. BOYDSTON, PETER A. KRENKEL

CONTENTS

1. Keynote Address: Putting Waste Heat in its Place 3
Jacob I. Bregman

2. Ecological Changes of Applied Significance Induced by the
Discharge of Heated Waters 15
H. A. Hawkes

Discussion 58
Eugene B. Welch

Discussion from the Floor 69

3. Water-Quality Standards for Temperature 72
Robert S. Burd

Discussion 78
Milo A. Churchill

Discussion 85
Carlos M. Fetterolf and Loring F. Oeming

Discussion from the Floor 92

4. The Cooling of Riverside Thermal-Power Plants 110
Andre Goubet

Discussion 124
John Erick Edinger

Discussion 133
G. Earl Harbeck, Jr.

5. Mechanics of Condenser-Water Discharge from Thermal-
Power Plants 144
Donald R. F. Harleman

Discussion 165
Norman H. Brooks

Discussion from the Floor 173

6. Modeling of Heated-Water Discharges 177
Peter Ackers

Discussion 213
Edward E. Driver and Rex A. Elder

Discussion from the Floor 222

7. The Horizontal Traveling Screen 225
Daniel W. Bates

8. Banquet Address: Thermal Pollution 243
Joe G. Moore

9. Design and Operation of Cooling Towers 249
Charles Waselkow

Discussion 272
Gordon L. Ford

10. Economic Considerations in Thermal Discharge to Streams 282
George O. G. Löf and John C. Ward

Discussion 302
W. R. Shade and Alex F. Smith III

Discussion from the Floor 308

11. Summary and Status of the Art 313
Frank L. Parker and Peter A. Krenkel

12. Research Needs for Thermal-Pollution Control 329
Bruce A. Tichenor and William A. Cawley

Discussion from the Floor 339

Participants in Informal Discussions 341

Index 343

LIST OF ILLUSTRATIONS

Chapter 2, Hawkes

Fig. 1. Diagram of an Ecosystem 18
Fig. 2. Temperature-Tolerance Ranges, Different Organisms .. 23
Fig. 3. Temperature-Tolerance Ranges of Various Fish 24

Discussion, Welch

Fig. 1. Paradise (Ky.) Steam Plant Surface Temperature, Auto-
trophic Index 60
Fig. 2. Paradise (Ky.) Steam Plant, 1966: Water Temperature,
Ammonia, Phosphate, Dissolved Oxygen 61
Fig. 3. Paradise (Ky.) Steam Plant, 1964: Zooplankton and
Water Temperature 62
Fig. 4. Wheeler Reservoir: Theoretical Temperature Below
Browns Ferry Nuclear Plant 64

Discussion, Churchill

Fig. 1. Temperatures Below Widows Creek Steam Plant, 1967 .. 80
Fig. 2. Temperature Distribution, Widows Creek Steam Plant,
1967 81
Fig. 3. Temperatures Below Colbert Steam Plant, 1967 82
Fig. 4. Temperature Distribution, Colbert Steam Plant, 1967 ... 82

Discussion, Fetterolf and Oeming

Fig. 1. Thermal Plume, J. H. Campbell Plant, Port Sheldon,
Michigan 89

Chapter 4, Goubet

Fig. 1. Temperatures Downstream from Beautor Power Station,
River Seine, 1966 115
Fig. 2. Temperatures Downstream from Vaires-sur-Marne Power
Station, River Seine, 1964 116
Fig. 3. Temperature Readings Below Power Plant at Vaires-sur-
Marne 117

Fig. 4. Thermal Section, River Marne 118

Discussion, Edinger

Fig. 1. Mechanisms of Heat Transfer Across a Water Surface . . 125
Fig. 2. Temperatures and Exchange Coefficients, Lake Colorado
        City, Texas 127
Fig. 3. Schematic Plan, Site No. 3 Cooling Lake 129
Fig. 4. Observed Temperatures, Site No. 3, 1966 129
Fig. 5. Longitudinal Profile of Temperatures, Site No. 3, 1966 . . 130
Fig. 6. Temperature Variations and Comparisons, Site No. 3,
        1966 130

Discussion, Harbeck

Fig. 1. Water Temperatures Below John Sevier Power Plant,
        Holston River 140
Fig. 2. Water Temperatures Below Shawville Power Plant, West
        Branch, Susquehanna River 141
Fig. 3. Compared Downstream Temperatures, Holston and Sus-
        quehanna Rivers 142

Chapter 5, Harleman

Fig. 1. Flow Stratification Near Power Plant 145
Fig. 2. Two-Layered Stratified Flow 146
Fig. 3. Cross-Section, Center Line of Outlet Channel 148
Fig. 4. Depth of Heated Layer Next to Outlet Channel 151
Fig. 5. River Interface Between Intake and Outlet Channels 152
Fig. 6. Cross-Section, Center Line of Intake Channel 153
Fig. 7. Interface Depth Next to Intake Channel 155
Fig. 8. River Interface Between Intake and Outlet Channels 155
Fig. 9. Heat Balance 158
Fig. 10. River Section at Power Plant Location 160
Fig. 11. Isotherms for Steady-State River Flows 161
Fig. 12. Downstream Temperature 162
Fig. 13. Critical Wedge Depth at Diffuser 162

Discussion, Brooks

Fig. 1. Submarine Outfall with Multiple-Port Diffuser 167
Fig. 2. Sewage Effluent Outfall, Whites Point, County Sanitation
        Districts, Los Angeles 167
Fig. 3. Buoyant-Jet Discharge into Homogeneous Fluid, Labo-
        ratory Tank 169
Fig. 4. Buoyant-Jet Discharge into Stratified Fluid, Laboratory
        Tank 169

Fig. 5. Density Data, Fig. 3 170
Fig. 6. Density Data, Fig. 4 170

Chapter 6, Ackers

Fig. 1. Existing and Prospective Power Stations, Severn Estuary . 183
Fig. 2. Variation of Specific Gravity with Temperature and
    Salinity 184
Fig. 3. Temperature Survey, Berkeley Power Station 186
Fig. 4. Longitudinal Temperature Profile, Berkeley Power
    Station 187
Fig. 5. Spread of a Buoyant Layer 188
Fig. 6. Black Rock Outfall-Intake Structure 190
Fig. 7. Hong Kong Harbor Power Stations 192
Fig. 8. Comparison, Field and Model Isotherms, Hong Kong . . . 193
Fig. 9. Outfall Structure, Hong Kong 195
Fig. 10. Hong Kong: Estimates of Recirculation on Spring Tide . 197
Fig. 11. Power Stations, Dublin Harbor 197
Fig. 12. Prototype and Model Salinity Profiles, Dublin 199
Fig. 13. Comparison, Model and Prototype Isotherms, High
    Water 200
Fig. 14. Comparison, Model and Prototype Isotherms, Low
    Water 201
Fig. 15. Tidal Forth 203
Fig. 16. Model-Prototype Comparison of Salinity Intrusion,
    Forth 204
Fig. 17. Heysham Harbor and Power Station Site 205
Fig. 18. Outfall Proposal for Heysham, with Thermal Survey . . 206

Discussion, Driver and Elder

Fig. 1. Cumberland Steam Plant 216
Fig. 2. Resistance Diagram 219
Fig. 3. 1:55 Model, Cumberland Steam Plant Condenser-Water
    Discharge Channel 220

Chapter 7, Bates

Fig. 1. Purpose of Work 225
Fig. 2. Leaburg Canal 226
Fig. 3. Industrial Water-Screen Design 227
Fig. 4. Industrial Water Screen 228
Fig. 5. Mayfield Louvers and Fish Bypass 229
Fig. 6. Mayfield Louvers 230
Fig. 7. Test Equipment 231

Fig. 8. Experimental Fish Screen 231
Fig. 9. Air Bubbles as Deflectors 232
Fig. 10. Advances in Fish Guidance Systems 232
Fig. 11. Design of the Traveling Screen 233
Fig. 12. Traveling Screen No. I 234
Fig. 13. Traveling Screen, End Turn 236
Fig. 14. Troy Test Flume, Grande Ronde River 237
Fig. 15. Traveling Screen, Artist's Concept 238
Fig. 16. Louver Structure at Tracy, California 239
Fig. 17. How Fish Avoid Impingement on Screen 240
Fig. 18. Traveling Screen at Power Plant Water Intake 241
Fig. 19. Traveling Screen at Angle to Flow 242

Chapter 9, Waselkow

Fig. 1. Cherokee (Colorado) Unit No. 4 Cooling Tower 250
Fig. 2. Cherokee (Colorado) Unit No. 4 Cooling Tower 251
Fig. 3. Two Man-Made Cooling Lakes, Valmont Station, Colo-
        rado 258
Fig. 4. Multi-Tower Installation on Common Tunnels, Nichols
        Station, Texas 261
Fig. 5. Damage to Main Cooling Towers, Zuni Station, Colorado,
        after 1965 Flood 262
Fig. 6. Undamaged Service Water Cooling Tower, Zuni Station,
        Colorado, after 1965 Flood 263
Fig. 7. Proposed Unit at Wyodak, Wyoming 264
Fig. 8. Relative Sizes of Wet and Dry Types of Cooling Towers . 265
Fig. 9. Natural-Draft Cooling Tower, Dry Type 266
Fig. 10. Fifty-Foot Coils of Dry Type of Natural-Draft Cooling
        Tower 267
Fig. 11. Possible Plant and Cooling-Tower Arrangement 268
Fig. 12. Typical GEA Installation 269
Fig. 13. GEA Air Condenser, End View 270
Fig. 14. Nuclear Generating Station, Fort St. Vrain, Colorado .. 271

Discussion, Ford

Fig. 1. Cooling-Water System 278
Fig. 2. Paradise (Ky.) Steam Plant and Cooling Towers 280

Chapter 10, Löf and Ward

Fig. 1. Water Use, Steam-Electric Utility Plants, 1959 285
Fig. 2. Annual Electricity Generation, Cooling Water Use, and
        Evaporation 287
Fig. 3. Efficiency and Water Use 288

Chapter 11, Parker and Krenkel
Fig. 1. 1900–1970 Growth in Use of Electric Energy and Selected Products 315
Fig. 2. Magnetohydrodynamic Power-Generation System 318

LIST OF TABLES

Chapter 2

Table 1. Thermal Springs Maximum Temperatures with Invertebrates Found 21

Table 2. Temperature Limits for Fish Survival, Death 33

Table 3. Upper Thermal Death Points and Temperature Preferenda of Some Fishes 37

Table 4. Summer Temperatures of British Rivers Receiving
Heated Discharges 43

Chapter 4

Table 1. Temperatures and Flow-Rate, River Seine, Near Montereau 119

Table 2. Measured and Calculated Temperatures, River Oise,
Near Beautor 121

Table 3. Measured and Calculated Temperatures, River Marne,
Near Vaires-sur-Marne 121

Chapter 6

Table 1. Inferred Reduction in Recirculation: Outfall Structure,
Hong Kong 194

Table 2. Comparison of Velocities in Outfall Jet 195

Table 3. Comparison of Maximum Velocities at Toe of Outfall
Apron 196

Table 4. Comparison of Recirculation Figures 196

Table 5. Comparison of Intake Temperatures, Dublin 199

Chapter 9

Table 1. Specification Figures: Wet Towers, Texas and Colorado 253

Chapter 10

Table 1. Values of K for Forced-Draft Cooling Towers 291

Chapter 11

    Table 1. Nuclear Power Costs 316
    Table 2. Types of Cooling Devices 326

ENGINEERING ASPECTS OF THERMAL POLLUTION

Chapter **1**   Keynote Address / Jacob I. Bregman

PUTTING WASTE HEAT IN ITS PLACE

I AM pleased to open this symposium on thermal pollution, the second of two such meetings to discuss this new and serious water pollution problem.

Vanderbilt University is a most appropriate spot for this meeting. This university is world renowned as an important center of inquiry and serious learning. The university and the Federal Water Pollution Control Administration of the Department of the Interior are jointly sponsoring these symposia. We are closer to effective management of this pollution problem because the resources of this great university, its talent and storehouse of knowledge, have zeroed in on the problem.

Secretary of the Interior Stewart Udall sends you his greetings and wishes for a successful symposium. We in the Department of the Interior will be following these meetings closely, for we hope for and expect from them new ideas and better solutions to help us meet the clean-water demands of this nation.

Salmon slosh belly-up onto the river banks; a foul odor wafts up from the estuary; green algae slimes the shores; and "No Swimming—Polluted," reads the sign posted on the beach.

You think some deadly chemical has done this work? No, it could be done by heat—waste heat—a valuable resource, but a resource out of place. Waste heat can cause serious thermal pollution, which can result in such disturbing water pollution effects.

"Resources out of place." That's what pollution has been called. When the resource out of place is heat wasted in industrial processes and transferred to great volumes of cooling-water—raising its temperature 10° to 20°—and dumped back into our waterways, you have *thermal* pollution.

3

This term *thermal pollution* means a substantial change of water temperature that adversely affects the water ecology and the life that depends on it. Let's not hide behind the more innocuous terms *thermal enrichment* or *thermal effects,* which sound more diplomatic but beg the question. Let's recognize the fact that heat can be a pollutant just as other misplaced resources can.

When water is raised above its normal temperature, the ecology of the waterway is altered because of the temperature increase, and, at the same time, the oxygen content is reduced, because less oxygen is held in solution at higher temperatures.

As a result, the aquatic life dies or changes, sometimes immediately, sometimes slowly; reduced spawning and hatching of eggs occur, along with disturbed life cycles and increased sensitivity to toxic materials. But thermal pollution is not just a fish problem. And its control is not just a question of "profits or pickerel."

The entire ecology of the river or estuary or coastal area can be altered by a change in temperature of just a few degrees. The fish and other aquatic life are part of the food chain, not just for river life, but for life on the adjacent lands and, eventually, for man himself.

Here are some examples of what can happen. An insect nymph in an artificially-warmed stream might emerge for its mating flight too early in the spring and be immobilized by the cold air.

A fish might hatch too early in the spring to find its natural food organisms because the food chain depends ultimately on the plants and these, in turn, depend upon day-length, as well as temperature. Fish, generally, depend on temperature changes in specific amounts to act as a signal for migration and spawning. The entire life cycle of fish may be upset by highly unnatural changes in the temperature cycle.

Trout eggs will not hatch if incubated in too-warm water, and salmon do not spawn if the temperature is too high. The sensitivity of all aquatic life to toxic substances is heightened at increased temperatures, and toxic effects of chemical substances are increased. Carp, for instance, are reported to be twice as susceptible to carbon dioxide in warm water as in water near the freezing point.

Biologists point out that the oxygen consumption by aquatic vertebrates doubles for every 10 degrees' rise in stream temperature. But as those temperatures rise, the water can hold less oxygen in solution. Thus, while supplies of dissolved oxygen steadily dwindle with increased temperatures, the demand for oxygen increases. Eventually, the aquatic life disappears.

As temperature rises, nuisance plants and rough fish flourish, while useful life dies. Recreation and drinking water deteriorate, as foul odors and algae slime appear.

Thermal pollution slows down the natural waste-assimilation rates of the waterway. So we spend more money for waste treatment, just to maintain the present quality of the waterway. If this extra treatment is not installed, pollution builds up, limiting all water uses.

Eventually, even further use as cooling water becomes limited when the water is too hot.

Not all thermal changes need be harmful. Some carefully controlled heat changes might permit new uses in certain locations. But these constructive uses of heated waters imply waste-heat management, so that waters can be heated just to the specific level required for the use and no more.

When we talk about thermal pollution today, we talk primarily about the thermal-electric power industry. This industry is the principal source of thermal pollution, and it is a controllable source. Thermal-electric power plants use 70 percent of all water withdrawn for industrial use for cooling and condensing.

To be sure, there are other sources of thermal pollution than power plants. But these are less significant sources. They are other industries— petroleum, chemical, steel, and pulp-and-paper processing—as well as water impounded in reservoirs and heated by the sun, and warmed irrigation-return flows.

Thermal power generation is the main source, because it wastes tremendous amounts of energy. The heat from fossil and nuclear fuel not converted into useful power goes mostly into cooling water which circulates through heat exchangers. For each kilowatt-hour of energy produced in a modern, highly efficient, coal-fired plant, 6000 BTUs, or about two-thirds of the heat, is waste and must be dissipated in cooling water. Nuclear plants are even less efficient, because of safety needs for lower-throttle steam temperature. Such plants waste as much as 40 percent more heat than fossil-fueled plants. For each kilowatt-hour produced in a nuclear reactor, 10,000 BTUs must be dissipated into cooling water.

Power is vital to the American economy and our high standard of living. So the power industry has had a phenomenal growth, to keep pace with the demands for commercial and residential power that have doubled about every ten years since World War II. But the growth of electricity is also beginning to mean the advent of thermal pollution—a type of pollution that has just begun to be recognized in the last year or two.

When power plants were relatively small, no one gave much thought to the problem of waste heat. But now they are big, numerous, and the trend is toward nuclear generation. All this means massive loads of waste heat.

By 1980, the power industry will be using a large portion of the total available freshwater runoff of the entire nation for cooling.

To meet future demands for power that will continue to double every ten years or less, 88 new nuclear-power plants are either being built or are in the planning stages. This compares to only 14 in operation today. Many more fossil-fueled plants will go into operation, too. But there will be a dramatic shift to nuclear-powered plants, which require 40 percent more cooling water than fossil-fueled plants.

Today, 95 percent of the thermally-generated electricity is still produced by fossil fuel, but this proportion is expected to decline to 65 percent by 1980.

Relatively non-polluting hydro-generation will slack off, as few hydro sites remain that are capable of economic development, and some of these are precluded from development by scenic and aesthetic reasons.

All power plants are getting bigger to achieve economies of scale.

The National Power Survey of the Federal Power Commission reports savings of 30 percent to 40 percent in production costs from a thermal plant of one million kilowatts compared to one of only 100,000 kilowatts. Regional pools will develop to take advantage of these economies of size, and of extra-high-voltage transmission lines, and will build larger-than-ever facilities for the production of power. The average size of all units retired between 1961 and 1965 was 22 megawatts. The average size of the 217 new fossil-fueled units under construction or in the planning stages for operation by 1971 is 295 megawatts.

The trend toward nuclear units means larger plants, still. The new nuclear units planned for operation by 1973 average 624 megawatts per unit.

As a result, larger, more concentrated loads of waste heat will be created and will have to be managed safely and efficiently for the protection of the environment.

As one example of what can happen, a proposed Vermont Yankee nuclear-power plant initially planned to use 60 percent of the maximum flow of the Connecticut River for cooling, raising the temperature 15° to 20°. As a result of public hostility, the plans have been changed to include cooling towers which cost several million dollars and which will drop the temperature increases to manageable amounts.

TECHNOLOGICAL CONTROLS

The problem can be controlled. We have technological and management solutions, although we know that they must be made more efficient and more economical. In fact, because the serious levels of waste heat will occur mostly in the future, thermal pollution is one form of pollution that we can prevent, rather than abate after it has fouled the waterways. We can do this if we plan to control thermal pollution from the very conception and design of new power plants.

Essentially, technological controls are of four types. Let us consider each one separately.

1) *Reduce heat wasted by improving the efficiency of thermal plants or developing new non-polluting energy systems.* Generating processes are more efficient today than they were 25 to 30 years ago, but they are still not good enough. Today, we use about 10,000 BTUs to produce one kwhr, on the average, compared with 16,500 BTUs per kwhr in 1938. Some highly efficient new plants are down to 8,900 BTUs per kwhr, but this is still only 38.2 percent efficiency. The Federal Power Commission suggests that average heat rates of 8,500 BTUs per kwhr are likely by 1980. While this would reduce thermal pollution by almost 20 percent, it still means only 40.2 percent efficiency. At best, then, by 1980, 60 percent of the heat from a coal-fired plant will be wasted and will have to be put to other uses or dissipated. The development of effective nuclear-breeder reactors should be an improvement over present nuclear systems, but they will continue to waste vast amounts of heat.

New and non-polluting methods of power generation are being intensively researched. High-temperature turbine-blade cooling could result in a decrease of as much as 50 percent of the cooling waters required. Electrogasdynamics, magnetohydrodynamics, and thermionic power generation, if they can be developed economically, could greatly reduce our water- and air-pollution problems. Fuel cells and thermal-electrical systems which do not require steam cycles for power generation also are being studied and present hope for the future.

2) *Manage the heat and the receiving environment to reduce the harmful impact of the heat.* These methods include increased turbulence; augmented river flow; better selection of sites for plants, from the point of view of ecological impact; or controlling the discharge of heat to maximize natural climate conditions.

3) *Dispose of excess heat through cooling towers, cooling ponds, or*

spray ponds. The evaporative cooling tower is probably the most widely used corrective device. There are two types: the inducted-draft type and the natural-draft type. The former has a slightly lower over-all cost than the latter. Evaporative cooling towers, however, put out a considerable amount of fog, and some types have high evaporation losses.

The capital costs and increased plant-operation costs of cooling towers are much too high. In addition, cooling towers themselves are a type of aesthetic pollution, since those required for large nuclear plants will rise as much as 30 stories high and be 2 city blocks in diameter. The Allegheny Power System has recently completed the two largest towers in the world, located in West Virginia. The two natural-draft towers are 370 feet high and 380 feet in diameter and serve one 540-megawatt unit each.

There is a danger, too, that cooling towers may topple over in a severe storm. This has already happened in Europe. This is not a pleasant prospect to consider, for a tower that may cost as much as two or three million dollars.

Summing up, there is a need for a vast improvement in cooling-tower design and a lowering of cost. Heat-transfer engineering is an old art, and I consider that vast improvements in cooling-tower technology can and must take place in the next few years. The financial stakes are high enough to warrant the research costs that it will take to do this.

4) *Make productive use of the huge amounts of waste heat.* I think this approach to the thermal-pollution problem is the one that is the most promising, not only for the power-generating community, but for all American industry. In the case of the electric utility, the productive use of the waste heat would help to make up for the relative thermal inefficiency of the steam-electric generating plant, as well as reduce the pollution problem.

Many uses of heat are currently being studied by industry and by the government. For example, in the Pacific Northwest, tests are being conducted to see if the heated water can be used for irrigation. Naturally, there have to be safeguards, because waters that are too hot or too cold may affect seedling emergence, plant growth rate, time of maturity, and crop yields. But with proper caution—management, again—it may be possible to extend the growing season and thus make constructive use of the waste heat.

Warm-water cultivation of oysters is being attempted in the East and in the state of Washington. What a treat for oyster lovers, if the experiments make it possible for oysters to spawn continuously for ten months a year and to reach their maturity in two and one-half years.

Thought is being given to warming waters along our northern beaches to add recreation opportunities for our growing population. And just think of the boost to the economy some of these areas would receive if their swimming seasons could be extended.

Utilities have been selling low-pressure steam for years to provide heat for buildings. Wider use of heat for this purpose would obviously be beneficial, all the way around.

Desalting is another important use that could be developed for waste heat, and the development of this approach is well under way.

Perhaps the most logical use for the waste heat is to plan for its conversion into an inexpensive energy source for satellite industries that will be constructed simultaneously with the power plant.

When there is no longer any question that the waste heat cannot just be put into the nearest body of water, then our American industrial ingenuity will certainly solve the problem of turning this waste heat into an economic asset that can be sold, rather than spend millions of dollars for cooling towers just to get rid of it.

Our industries are based on the use of energy; vast sums of money are spent for it. Why not surround the power plant with new industry and bring about the development of nuclear-industrial parks? Why not computerize the planning of entire industrial complexes based on a nuclear-power center instead of our present disorganized approach to industrial siting? We may stand on the brink of a modern industrial revolution which could be triggered by the absurd fact that fish can't stand heat.

WATER-QUALITY STANDARDS

All these controls, in all four categories, are most effective, most efficient, and cheaper to build into a plant from initial design stages than to add on to a plant after it is built. But why build them in at all, if an industry is not particularly concerned with the environment?

Why, because we have legal incentives to installing waste-heat controls such as these. Today, we have water-quality standards: the basic approach for controlling all water pollution. The Water Quality Act of 1965 required that water-quality standards be set for all interstate and coastal waters, first, by the states: then, to be approved by the federal government, to become federal standards, too. A water-quality standard consists of two parts:

1) Water-quality criteria which are scientific limits for heat, dissolved oxygen, acidity, and so on, that permit desired and beneficial uses of the

waters, such as recreation, fishing, or municipal usage, agricultural or industrial supply; and

2) A plan to implement and enforce these criteria, including a schedule for remedial measures needed to achieve water of the assigned quality, defined, city-by-city and industry-by-industry, as well as plans to enforce the standards legally.

The specific limit for heat set in the standards generally sets a mixing zone: the area where the heated effluent can join the receiving waters. It imposes maximum-temperature limits for a waterway and prescribes maximum-temperature deviations above or below the normal temperature, as well as limiting the duration of that deviation.

Forty-one of fifty-five sets of standards have been approved, in whole or in part. We expect that the rest will be approved soon.

In considering water-quality criteria, Secretary Udall has taken the approach that, where exact limits are not known, numerical values will be set that err on the side of safety. Thus, if there is an argument over whether certain types of fish can survive at 93° F., but everyone agrees that they can tolerate 90° F., the criterion is set at 90° F., until such time as more evidence will indicate that we can safely go higher. The same approach has been taken with regard to allowable increase above background. The ecology of a stream will not be destroyed in an experiment to see if the fish can stand the added temperature. Rather, the ecology will be maintained within accepted safe limits, and those limits will then be researched to find out if they can be expanded safely.

ENFORCING WATER-QUALITY STANDARDS

Both the criteria of water quality—temperature limits, for example—and the timetables for remedial measures are enforceable by the states and by the federal government in the courts. But there is a problem: the Federal Water Pollution Control Act authorizes the Department of the Interior to enforce the standards only *after* a violation has occurred. In other words, a power plant would have to be built and operating and discharging heat to raise the stream temperature above the limit set in the standards before we could go in and legally abate that pollution, which usually would mean shutting down the plant until it installs corrective measures.

This is certainly not the most effective or efficient approach to standards-enforcement or to water-quality management, either from industry's or government's point of view.

In fact, at this time, no formal attention is given by any federal agency to regulating construction of thermal power plants, from the standpoint of thermal pollution, *before* that construction occurs.

The Atomic Energy Commission has said that its licensing law confines it to considering only the question of radiological health and safety, not thermal pollution; and the AEC does not believe the Federal Executive Order on water-pollution control gives it the needed authority to regulate construction in terms of thermal pollution. The Department of Justice has agreed with these views.

The Federal Power Commission is directed by its law to consider the effect of its license on various water uses—in effect, water quality—but the FPC does not license fossil-fuel plants, only hydroelectric plants.

But despite the absence of "pre-construction authority," the Department of the Interior is urging the utility industry to work with the FWPCA and the appropriate state officials during the planning and design stages of new thermal projects, so that adequate safeguards for the protection of water can be built into the project.

This is very constructive, but still informal and does not, unfortunately, prevail throughout the industry.

A formal "pre-construction review" to assure that thermal plants comply with water-quality standards is what is needed. It saves money, time, and a lot of industry and government headaches. It permits adequate controls to be built in from the start, rather than added on after construction of a plant that violates water-quality standards and causes an enforcement suit.

There are several proposals in the Congress for new legislation to create pre-construction-review authority. Essentially, these propose one of three things:

1) Review of thermal-standards compliance by the licensing agency; or

2) The Department of the Interior's recommending, in an advisory capacity to the licensing agency, the impact of a license on standards; or

3) Certification of compliance by the Department of the Interior.

LICENSING AGENCY REVIEW

The first proposal for pre-construction review of thermal-electric plants— and, more particularly, atomic-fired plants—is one made by Senator Edmund Muskie and included as Section 5 of S.3206 which passed the Senate on July 11, 1968, and is now pending in the House Public Works Committee.

This legislation would extend to any federal agency the authority necessary, plus a directive to consider compliance with water-quality standards in contracts, permits, licenses, and leases.

During floor debate, Senator Muskie said: "The amendment to Section 11 is intended to clarify the position of the Congress in this regard by imposing a requirement that federal agencies cooperate with pollution-control efforts of the Secretary of the Interior."

ADVISORY ROLE FOR INTERIOR

The second route being proposed for pre-construction review is somewhat more formal. It would expand the power of the AEC to consider thermal pollution in its licensing and combine this with an advisory role for Interior. The final decision on the effect of the licensed facility on standards would be left to the AEC. Proposals along these lines are being considered by the Joint Committee on Atomic Energy.

The problem is that this proposal gives the AEC jurisdiction to make the final judgment on water quality. The AEC could then license an activity which another agency—the Department of Interior, for instance—might consider a violation of the pollution laws, causing an enforcement case. This would be a very real problem for government and industry. In addition, the power-regulatory agencies do not have the pollution-control expertise. To get that expertise means duplicating staff and talents. Even with identical experts, the power agencies have essentially a conflicting resource mission in the case of thermal pollution, and there is no reason why water quality must take a back seat to power production.

Since thermal control is primarily a pollution problem, it should be under the jurisdiction of the pollution-control agency.

DEPARTMENT OF THE INTERIOR CERTIFICATION

Certification—the third route—would require that the Atomic Energy Commission, or any federal agency, before issuing a permit or license, obtain a certification from Interior, stating that the heat to be discharged would not violate the applicable water-quality standards.

In effect, Interior would have the last word on effects of a license on standards. Legislation introduced by Representative John Dingell of Michigan would create this authority.

There are arguments against certification, the main one being that there are delays when two agencies are involved in a construction permit. But there are also very real benefits to the nation and the industry from certification.

First, the nation benefits, because certification of water quality assures that clean-water interests of the nation are considered in the licensing procedure, in addition to power interests.

Second, the utilities benefit from certification. Any utility desiring to build a thermal power plant is entitled to know what Interior's position is, and will be, regarding the plant's effect on water-quality standards, to prevent litigation on water-quality standards after construction of a plant. The weight of evidence in a water-quality standards-enforcement suit would be with the industry that had prior certification, and so such cases should not arise.

Certification saves a company money. Cooling facilities, added on to a plant after it is built, and standards which are violated are much more expensive and less efficient than those designed and built in from the start. Or, worse, what if adequate cooling facilities cannot be added on, after construction, to compensate for a basically faulty site, from an environment standpoint?

Certification by Interior would provide technical aid to industry. Interior could extend its advice early on sites and controls, even before the industry applied to the licensing agency. Specific devices and sites could be recommended. This would be like an industry getting the opinion of the Justice Department prior to a merger, to see in advance the legal consequences of that merger.

The question of dual permits need not arise, since Interior certification could be made a part of the over-all licensing procedure, with the AEC permit not being granted without a showing of an Interior certification.

In summary, let me say that I think this conference can be very useful in our efforts to understand and cope with thermal pollution. There are many things we do not know about this complex environmental change. While we know that serious changes do occur from heat, we do not have enough specific information about incremental changes and effects of various rates of change. We must also develop cheaper and more efficient technology to manage the waste heat.

Let me say that the question, though, for our consideration is *how* we control thermal pollution, not *if* we control it. Laws on the books today have answered that; and proposed stronger laws, too, would require even more effective controls.

As long as power plants must have controls on waste heat, we should make them as cheap and efficient as possible. This means that we should build them into thermal power plants from the initial design stages: we should program waste-heat management into every phase.

There are two approaches to heat management that I would particularly like to see this conference explore and emphasize: one, we need to find ways to reduce the production of waste heat; and two, we need to develop new and more profitable ways of putting to work the waste heat that cannot be eliminated.

What these approaches add up to is putting our heat resource back where it belongs, back into the useful cycle. Let's put our waste heat in its place.

Chapter 2   H. A. Hawkes

ECOLOGICAL CHANGES OF APPLIED SIGNIFICANCE INDUCED BY THE DISCHARGE OF HEATED WATERS

THE increase in temperature of surface waters resulting from the discharge of heated water affects water quality in two ways: directly and indirectly. The direct effect of the raised temperature may be detrimental to man's interests if, for example, it is to be used for further cooling processes, or if required as a potable water supply. In some cases, the raised temperatures may be beneficial—for recreational bathing, for example, or, probably of greater economic significance, for the keeping open of navigable rivers throughout the winter period when normally they would be frozen. The indirect or secondary effects induced by increased temperature, involving ecological changes in the biology and chemistry, may also be of applied significance. The aim of this contribution is to consider the applied significance of these induced ecological changes.

With the increasing awareness of the problems of thermal pollution, some ecological surveys have been carried out on affected waters. Data from these surveys is now becoming available, and I understand the first part of this symposium dealt with such biological and chemical aspects of the problem. Where organisms of economic or recreational interest, such as fish, are affected, the applied significance is evident. Where, however, "lesser lights" are involved, as, for example, the reported increase in the abundance of the diatom *Navicula ambigua* (Stangenberg and Pawlaczyk, 1960), an increase in the population of the snail *Physa acuta,* the appearance of a tropical worm *Branchiura sowerbyi* (Mann, 1965) or the disappearance of the flatworm *Dugesia lugubris* (Mann, 1965), the applied significance is less obvious. Even the more biologically-minded American sanitary engineer may join his British civil-engineer counterpart in commenting "So what?" to such results. A secondary aim of this paper is, therefore, to attempt to answer this very question. In order to achieve these objectives, it is first necessary to establish certain fundamental ecological principles and then consider the importance of temperature as a factor determining the structure and function of aquatic communities. On the basis of these fundamental principles, the ways in which induced ecological changes may be of applied significance are considered. Results of laboratory experimental work and field surveys will be reviewed in assessing the degree of change induced by thermal discharges.

RELEVANT ECOLOGICAL PRINCIPLES

To understand fully the effects of increased temperatures on the biology and chemistry of fresh waters, it is first necessary to have some understanding of the ecology of fresh waters involving the interaction of chemical and biotic factors.

Water as a Medium for Life

Compared with the influence of the atmosphere on subaerial organisms, water as a medium has a greater influence, affecting both the structure and physiology of aquatic organisms. Water contains many of the materials essential to life, in limited amounts, and consequently the activity of organisms may bring about substantial changes in the concentrations of these substances in the water. Thus, aquatic organisms and the water in which they live coexist in a very intimate relationship. The resultant mutual interaction between aquatic organisms and the water is of vital significance in many aspects of water management involving water storage, treatment, waste-water treatment, and water-quality control. The community of organisms supported by a water depends in part upon the quality or nature of the water; and, conversely, the quality of the water for many purposes is affected by the types of organisms present and their activity.

In the natural hydrologic cycle, water falling on the earth's surface already contains gases, such as oxygen, carbon dioxide, and oxides of nitrogen, dissolved from the atmosphere. It may also contain micro-organisms, such as bacteria, fungi, and algae, which are carried into the atmosphere by air currents from the land. Thus, natural water arriving on the earth's surface is neither pure in a chemical sense (H_2O) nor is it lifeless or sterile microbiologically. Water which percolates into the ground has its microbial population reduced and becomes isolated from the influence of the sun's energy and is therefore relatively biologically inert and is usually of good quality. Water flowing over the surface of the earth collects more organisms and nutrient materials, which pass into rivers and lakes. Such surface waters, containing biogenic substances—minerals and essential gases—inoculated and subjected to sunlight, are potentially biologically active and support an active ecosystem.

Concept of a Functional Ecosystem

An ecosystem is an ecological system in which there is an interaction between the living organisms and the nonliving (abiotic) environments to produce a cycling of material and a transfer of energy as illustrated in Figure 1. Ecosystems usually have three component groups of organisms: producers, consumers, and decomposers.

Producers.—These utilize light energy to synthesize organic matter from inorganic sources: carbon dioxide, water and mineral salts. In fresh waters, producers are represented by algae and aquatic vegetation.

Consumers.—These are organisms in the community, mostly animals, which use the producers as their food either directly as *Herbivores (primary consumers)* or via food chains, as *carnivores (secondary or tertiary consumers).*

Decomposers.—Waste products and corpses from the consumers, together with dead vegetable matter and cells from the producers, accumulate as dead organic matter to form detritus. This dead organic matter is used as food by the third component of the ecosystem, the *decomposers*. These are prominently the bacteria, fungi, and saprozoic protozoa assisted by scavenging animals. Most decomposers using organic food are therefore Heterotrophic; but the final stage of mineralization are carried out by Autotrophic bacteria, e.g., nitrifying bacteria. This mineralization completes the cycle by making mineral nutrients again available to the producers.

In fresh waters, therefore, as in all natural habitats, organic matter and organisms responsible for its breakdown form an essential part of the ecosystem.

It follows from this concept of an ecosystem, involving the interrelationship of organisms between different groups and between them and their abiotic environment, that in considering the effects of any factor, such as temperature, the over-all effect on the species within the ecosystem must

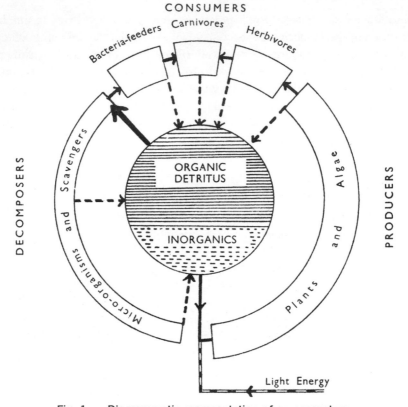

Fig. 1——. Diagrammatic representation of an ecosystem.

Reproduced from "Some Ecological Aspects of Water Conservation," by H. A. Hawkes, in **Proceedings of Symposium Conservation and Reclamation of Water,** Institute of Water Pollution Control, 1967.

be considered (synecology), as well as the direct effect on the species (autecology). Thus, in considering the effect of increased temperatures on fish, for example, it is necessary not only to consider the tolerance range of the fish, but the effects on organisms in the fish-food chain and on the abiotic component, such as dissolved oxygen.

Autecology

Autecology is the ecology of individual species, involving studies of factors influencing their distribution in nature. Synecology, on the other hand, deals with communities and the interrelationships of populations within them. In studies on the effects of heated waters, we are concerned with both these aspects of ecology. An organism is affected by a set of ecological conditions. For an organism to thrive, all conditions must be favorable. If any one factor becomes unfavorable and approaches the limits of tolerance for the species, it becomes the limiting factor restricting the distribution of that species. This is known as Skelford's law of tolerance (Odum, 1959). Each species has its own tolerance range for different factors. Species may have a wide tolerance range for one factor and a narrow range for another factor. Species with a wide tolerance range for temperature, for example, are said to be *Eurythermal;* those with a narrow range are *Stenothermal* (Figure 2). In spite of Skelford's law, factors may interact; for example, the effect of increasing temperature above the optimum is influenced by the concentration of dissolved oxygen. Furthermore, a given species may have different tolerance ranges and optima at different stages in its life history. Such a sequence of changes is provided by the normal seasonal climatic changes in temperate zones.

Although tolerance ecology goes a long way to explain the distribution of a given species, it is often found that an organism is absent when all the autecological factors are optimal. Also, organisms are often found to be most abundant towards the fringe of their tolerance range. In other cases, a species is more restricted in its distribution in nature than required by the tolerance ranges. An explanation of these phenomena is to be found in synecological terms. Factors usually operate not on a single population but by affecting the outcome of competing populations.

TEMPERATURE AS AN ECOLOGICAL FACTOR

Unlike the higher organisms, such as birds and mammals, which have evolved a most efficient self-regulatory mechanism for controlling their body temperature, the body temperature of other organisms fluctuates with that of the external environment. Such organisms are said to be poikilothermal.

Ecological Temperature Range in Nature

Freshwater organisms as a group occupy a wide range of the temperatures available between freezing and boiling points. Micro-organisms occupy the upper and lower extremes. Some algae are known to photosynthesize actively at temperatures approaching freezing, below ice in lakes. At the upper temperature extreme, the organisms found in hot springs provide evidence of thermal tolerance. In the hottest springs, no life was found; but bacteria were found in water at 85° to 88° C. (185° to 190.4° F.)[1] and

1. Temperature equivalents given throughout in parentheses are only approximate figures.

algae at 73° C. (163.4° F.) (Brock, 1966). Odum (1959) quotes the following temperature tolerances for micro-organisms in hot springs:

Bacteria	(88° C.)	190° F.
Cyanophyceae (Blue-green algae)	(80° C.)	176° F.
Protozoa	(54° C.)	129° F.

Brues (1939) carried out a survey of thermal springs in the Dutch East Indies. Table 1 indicates the maximum temperature tolerance quoted in his results for different invertebrate groups. Apart from high temperatures, many of these thermal springs are characterized by high salinity and abnormal pH values.

Physiological Considerations

All organisms depend upon metabolic respiration for their existence and activity. Respiration involves chemical reactions and is therefore temperature-dependent, the reaction velocity increasing with increasing temperature. These biochemical reactions are, however, catalyzed by enzymes which are themselves sensitive to temperature changes, being inactivated at higher temperatures. Thus, above a certain temperature, the optimum, the thermal inactivation of enzymes more than offsets the increase in reaction velocity and results in a rapid decline in over-all activity and, ultimately, in death.

The effect of thermal inactivation is a function of time, and as pointed out by Baldwin (1959) in determining the optimum temperature, the time factor must be taken into account. In tests of short duration, during which the enzymes do not need to be long-lived, much higher "optimum" temperatures may be recorded than in tests of longer duration in which the effects of enzyme inactivation become evident. The same may apply in laboratory studies to determine upper lethal temperatures.

In the case of micro-organisms, the optimum of the thermal range is usually taken as the temperature at which most rapid increase in population of a species occurs. Increase in temperature may, at certain temperature levels, cause momentary increase in growth but at the same time produce slow, irreversible injury. The "limiting optimum" is the highest temperature at which most rapid increase occurs which is not followed by decline due to irreversible inhibition (Hutchinson, 1967).

Thermal Death.—Most enzymes are only inactivated by temperature above 35° C. (95° F.) and thus, although thermal death at high temperatures, especially in the case of micro-organisms, may be accounted for by

TABLE 1. Maximum temperatures at which representatives of different
invertebrate groups were found in Thermal Springs.

Systematic Group	Species	Maximum temperature at which found. approximate equivalent	
		°C	°F
Protozoa	**Nebala** spp.	45	113
Nematoda	Several spp.	48.5	119
	Many spp.	30—45	86 —113
Annelida			
Oligochaeta			
Enchytraeidae	**Enchytraeus spp.**	41	105.8
Hirudinea	**Limnatis manillensis**	38	100.4
Arthropoda			
Crustacea			
Cladocera	2 Spp.	35.5	95.7
		34	93.2
Copepoda	Several Spp.	34—42	93.2—107.6
Insecta			
Odonata	5 Spp. of subfamily Libellutinae	36—38	96.8—100.4
Hemiptera	Several Spp. from 10 families	32—38	89.6—100.1
Diptera			
Chironomidae	**Chironomus inferior**	40	104
	Tanypus punctipennis	39	102.2
	Several Spp. of Chironominae	34—38	93.2—100.4
Ceratopogonidae	**Bezzia assimilis**	39	102.2
Calicidae	**Anopheles spp.**	35—38	95 —100.4
Stratiomidae	**Odontymyia sp.**	35—37	95 — 98.6
Ephydridae	Several spp. including genera **Ephydra** Scatella	39	102.2
Mollusca	**Lymnaea spp.**	37	98.6
	Viviparus costatus	37	98.6

SOURCE: From data quoted by Brues, 1939.

thermal inactivation of the enzyme systems, the upper thermal limits of many freshwater organisms, many of which are below 20° C. (68° F.), must involve other physiological processes. Furthermore, the different tolerant ranges of different organisms and even of different stages of the same organisms would mean they each possessed quite different enzymatic systems, if thermal inactivation of enzymes was the only process involved.

It is probable that different systems are affected in different groups of organisms. Due to the increased oxygen demand in the respiring cells at higher temperatures and limitations in the rate of oxygen transport to the cells, oxygen deficiency in the cell would be caused, with resultant harm. The eggs of salmonids have much lower upper-thermal limits than the newly hatched alevins. This could be due to the presence of the egg membrane, which slows down the transport of oxygen. Aquatic organisms are generally more sensitive to high temperatures than corresponding subaerial species. If the availability of oxygen is involved in determining upper thermal limits, then the reduced availability of oxygen as dissolved oxygen in water could account for the difference between aquatic and subaerial organisms in relation to thermal tolerance. In the higher animals, adverse effects on the co-ordinating mechanisms of the central nervous system or on "pace-making" mechanisms may be involved. The melting points of body fats may also be a factor determining the upper lethal temperature. Heilbrunn (1955) considers that changes in the lipid structures of the cell membrane are responsible.

Many freshwater organisms, in contrast to marine forms, are faced with the problem of maintaining the ion concentration of their internal body fluids at a higher concentration than the external water. This they achieve by the process of osmo-regulation. At higher temperatures, although the permeability of the cell membranes is increased, the physiological processes of osmo-regulation are more efficient.

Adaptation.—Although the body temperatures of poikilotherms fluctuate with the external temperature of the water, some, at least, exhibit a rate of activity which is independent of temperature over a wide range (Bullock, 1955). This he considered to represent a degree of compensation to changes in temperature. It is now well established that many freshwater organisms, especially fish, are able to tolerate higher temperatures experimentally by first "acclimating". For example, by increasing the acclimating temperature by 3° C., the lethal temperature for roach *(Rutilus rutilus)* is raised by 1° C. (Cocking, 1959). This occurs over much of its thermal range. Eventually, the acclimation temperature reaches the lethal temperature and this is

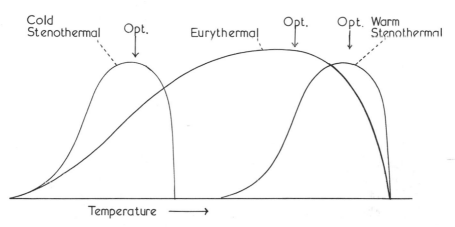

Fig. 2—. Diagram illustrating the differences in temperature-tolerance ranges of organisms.

known as the ultimate upper lethal temperature. Acclimation is usually a rapid process—less than one day—and is only slowly lost. The phenomenon of acclimation is evidence of physiological adaptation and may be of ecological significance as survival value in nature. Slower adaptation under natural conditions known as acclimatization is also known to occur. Within a single species, genetic differences in thermal tolerance are known.

The possibility of evolutionary adaptation to higher temperatures by genetic changes is demonstrated by the discovery by Banta and Wood (1928) of a mutation in the freshwater crustacean *Daphnia laevis*. The original strain of this species was found to be cultured best between 18° and 20° C. and was only raised with difficulty at the limits of its range, 27° to 29° C. Then a thermal clone arose, best reared at 27° to 29° C., which did very poorly below 20° C.

Temperature as a Factor Influencing Distribution and Seasonal Incidence of Organisms

Whatever the physiological causes, different species exhibit different temperature-tolerance ranges and are confined in nature within different temperature zones. This is represented diagrammatically in Figure 3. Organisms tolerant of a wide range of temperature are said to be eurythermal; those with a narrow range, stenothermal. Stenotherms in low temperatures are referred to as oligothermal, those in warm waters as polythermal. These

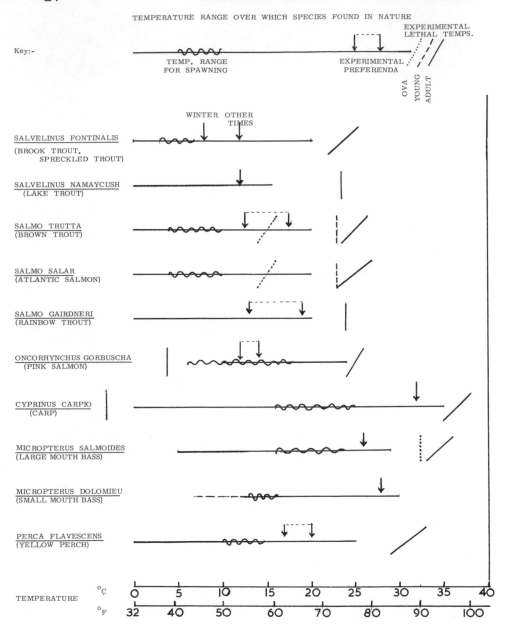

Fig. 3—. Temperature-tolerance ranges of various fish.
SOURCE: From data after various authors quoted in Wurtz and Renn, 1964.

temperature requirements are important in determining the distribution of organisms in fresh waters.

Distribution of Species in Streams

The complex way in which temperature may operate as a factor influencing distribution is well illustrated by considering the distribution of certain Ephemeroptera (mayflies). Ide (1935) found, by examining a Canadian river in which there were marked differences in temperature between the successive stations, that there was an increase in the numbers of species of mayflies downstream associated with the greater fluctuations in temperature. Two species found near the source, however, disappeared downstream. This was attributed to the higher summer temperatures at the lower stations. The absence of many species from the upper stations was attributed to the fact that the temperature of the water did not rise sufficiently to allow them to grow and complete their development. Macan (1963) found that the mayfly nymph *Heptagenia lateralis* was abundant in streams where the highest temperature was about 18° C. (64.4° F.) or lower. It was absent or rare in most other streams where the summer temperature rose above 18° C. (64.4° F.). This species cannot develop at low temperatures. It was suggested that its absence from the warmer becks was due to the fact that the temperature rose from the threshold temperature for development to the lethal temperature in a time shorter than the development period for the species. The same species is present in Lake Windermere, where the summer temperature may exceed 20° C. (68° F.). However, the lake warms up more slowly than the becks, and the lethal temperature is not reached until after the mayflies have emerged. The eggs are resistant to the higher temperatures. The distribution of *Prosimulium* larvae in hill streams may be similarly explained (Davies and Smith, 1958). Thus, probably four critical temperatures need to be satisfied for the successful life history of such insects: (1) sublethal temperature for eggs, (2) the threshold temperature for development, (3) the lethal temperature for nymphs, and (4) suitable temperature for emergence and mating flight.

In nature, a species is often found to be more restricted in its distribution than its thermal limitations would demand. The flatworm *Crenobia alpina* is rarely found in streams in Britain at temperatures above 15° C. (59° F.) but it has been found to withstand short periods of exposure to 25° C. (77° F.) and longer periods at 20° C. (68° F.) in laboratory tests (Schlieper and Bläsing, 1952).

Seasonal fluctuations in temperature are of ecological significance. Besides providing the different optimum temperatures needed from different developmental stages, they permit a series of different species to become successively established throughout the year, thus enabling several species to occupy the one habitat, so increasing the complexity of the community and thereby strengthening its stability. Such seasonal changes are most significant in lakes in relation to plankton successions.

Seasonal Successions in Lake Plankton

Seasonal successions in plankton species in lakes are best considered in terms of synecology. Changing temperature favors a succession of species in inter-specific competition, depending upon their temperature optima. This is most evident in the case of zooplankton.

Zooplankton.—Crustacea increase their rate of multiplication with rising temperatures over their respective tolerant range. Furthermore, the increase in multiplication rate with increasing temperature is greater in the case of the warm-water species.

In case of some crustacea, autecological factors may themselves operate to restrict plankton species. *Limnocalanus macrurus,* a cold stenothermal copepod which deposits its eggs only when the temperature is below 7° C. (44.6° F.) is therefore a winter species in many lakes; but in the summer it may be found in the hypolimnion of other lakes in which it thereby breeds throughout the year (Hutchinson, 1967).

Phytoplankton.—Hutchinson (1967) discusses in detail the factors affecting the seasonal cycle of phytoplankton in lakes. In considering temperature as a factor, he recognizes three temperature ranges in relation to each species: sub-optimal, optimal and supra-optimal. At sub-optimal values, an initially low population will increase less rapidly than when conditions are optimal. However, even at sub-optimal temperatures, an established large population may persist and even increase with decreasing values of temperature below the optimum. At supra-optimal temperatures, active inhibition and death result. Hutchinson (1967) quotes the following approximate thermal optima for various plankton algae:

Melosira islandica helvetica	5° (41° F.)
Synura uvella	5° (41° F.)
Asterionella formosa	10° to 20° C. (50° to 68° F.)
Fragilaria crotonensis	15° C. (59° F.)
Ankistrodesmus falcatus	25° C. (77.6° F.)
Scenedesmus ⎫	
Chlorella ⎪	
Pediastrum ⎬	20° to 25° C. (50° to 77.6° F.)
Coelastrum ⎭	

Some species are reported to form thermal races or ecotypes. In lakes in the Austrian Alps, Ruttner (1957) considers two races of the *Asterio-*

nella formosa exist. One exhibits stenothermal characters having an optimum temperature for development between 5.1° and 8.0° C. whilst the other is more eurythermal and has its optimum about 12.2° C. Many field observations of algal seasonal successions suggest they are temperature-induced and imply that some species, such as the blue-green algae, *Oscillatoria rubescens,* are cold-water forms, whilst other blue-greens, including *Anabaena flos-aquae* and the dinoflagellate *Ceratium hirundinella* and the desmid *Staurastrum,* are warm-water species.

If such successions are thermally determined, then one would expect the same mesothermal species to be dominant at equivalent temperatures in spring and autumn. Early investigations (Wesenberg-Lund) found this to be the case with the diatom *Fragilaria crotonensis* in Danish lakes. The diatom had two periods of abundance, one in the spring and the other in the autumn, being replaced by blue-green algae in the summer at temperatures of 22° C. (71.6° F.). In the cold summer of 1902, however, the water temperature did not exceed 16° C. (60.8° F.) and the diatom persisted throughout the summer, suggesting that temperature itself was the main cause of the succession in this case. However, later work has shown that *Fragilaria crotonensis* is an exception.

Although temperature is probably an important factor controlling both the abundance and species composition of the phytoplankton, other factors associated with seasonal changes, such as light, may also be involved. The spring increase in algal growths is often associated with increase in light intensity, rather than increases in temperature. British workers tend to stress the importance of nutrients (Pearsall, 1923). Seasonal changes in the availability of nutrients, especially in lakes which stratify, may also be a contributory factor. High nutrients in the spring, it is suggested, cause the vernal diatom activity, *Asterionella formosa,* which requires a higher nutrient level appearing before *Tabellaria fenestrata.* At the end of spring, when the silicate concentration has been reduced and when the $NO_3 : PO_4$ increases, *Dinobryon divergens* often replaces the diatoms. At low-nutrient conditions in the summer, the green algae, including the desmids, occur. Subsequent studies (Lund, 1950) have demonstrated that the incidence of the diatom *Asterionella formosa* in Windermere was due to variations in the dissolved-silica concentration, the decline in the population occurring when the silica concentration fell below 0.5 mg/1. It would be expected that the availability of nutrients such as N. and P. would limit the population of algae in a similar way. However, as discussed by Fogg (1965), there is little conclusive evidence of this from field studies.

Other seasonal changes which need to be considered in accounting for seasonal phytoplankton successions are turbulence, accessory nutrients, vitamins, inhibitory substances, predatism, and parasitism.

It would thus be unwise, in predicting the effect of raising lake temperatures by thermal discharges, to assume that the natural seasonal successions in lakes were entirely temperature-regulated.

POSSIBLE CHANGES OF APPLIED SIGNIFICANCE CAUSED BY INCREASED TEMPERATURES

The foregoing account demonstrates the importance of temperature as an ecological factor affecting the activity, distribution, incidence, and abundance of aquatic organisms. It is therefore to be expected that increasing the water temperature will have marked effects on the biota. At this stage, it is necessary to consider which of the groups of organisms affected are of economic or applied significance as a basis for detailed consideration. The possible effects on these of increased temperatures will then be considered in light of their known temperature requirements from field and laboratory observations. An appreciation of the complexity of ways by which the temperature factor operates in aquatic ecosystems, outlined earlier, makes one realize the difficulty of predicting the change in the biota quantitatively. Temperature tolerance ranges and preferenda determined by laboratory tests, although useful, are not in themselves sufficient to predict the over-all effect of increased temperatures. The limited available data from surveys of waters receiving thermal discharges will also be reviewed.

Because of their recreational and economic value, fish are often considered to be the most important aquatic organisms. Fish, however, need to feed, many of them on aquatic invertebrates. Invertebrates may also be of significance in their own right, as nuisance organisms—e.g., midges— the larval stages of many of which are aquatic. Other invertebrates are of medical or veterinary importance as vectors of diseases of man and his cattle, e.g., mosquitoes, *Anopheles* and *Culex,* the buffalo gnat *Simulium,* all of which have aquatic stages, and certain freshwater snails, vectors of blood flukes *Schistosoma.*

Although of less evidence to the general public, except the few causing disease, micro-organisms are essential for the welfare of aquatic ecosystems, and hence the effects of increased temperatures on their activities is of fundamental significance. This is especially true of waters suffering organic and toxic pollution. The profuse development of some micro-organisms may cause nuisance: e.g., slime infestations and excessive growth of filamentous algae *(Cladophora)* in rivers, algal blooms in lakes, and the development of nuisance algae in water supplies. Weed infestations in rivers and lakes is a further problem which may be affected by increased temperatures.

Finally, we come to what I understand is a vexed question, that of the value of the aquatic biota as indicating water quality. I have previously expressed my views on this subject in several publications (Hawkes, 1962, 1963, 1964). Realizing that British papers in this field of study are generally neglected in America, I may take this opportunity of summarizing these views, since they are relevant to the present discussion.

Significances of Induced Changes in the Biota in Assessing Water Quality.—From numerous surveys throughout the world, the effects of different types of discharge on stream biota have been established. The application of such findings, however, in assessing pollution has given rise to much controversy. The difficulty arises in defining pollution: what degree of change constitutes pollution? This difficulty also applies to other methods of assessment.

Concepts of Pollution.—River pollution has been defined as the addition of something to water which changes its natural qualities. If stream water were H_2O, its "natural qualities" could be precisely defined and any changes readily measured by physical and chemical tests. It is, however, a dilute nutrient solution supporting a community of living organisms. Furthermore, its composition may differ in nature with time and locality. Changes in the natural qualities of river water are sometimes difficult to detect and measure chemically but are usually clearly reflected by changes in the stream community. On this concept, pollution can be readily detected and assessed by biological examination. Indeed, many biologists accept the concept of pollution as implying an upset in the balance of stream communities.

Other workers would define pollution in more practical terms and contend that pollution involves the impairment of the potential uses of the water. A river, however, is more than a flowing mass of water, and river pollution implies more than water pollution. River pollution can, then, be defined as the discharge of something to the river which so changes its nature that the general amenities of the river are adversely affected, its suitability for man's legitimate uses being impaired. These would normally include utilitarian uses, such as navigation, industrial, agricultural, and domestic requirements; recreational uses, including fishing, boating, and bathing; and also the satisfaction of man's aesthetic interests.

The acceptance of this more practical concept of pollution necessitates a re-assessment of the validity of biological methods. No longer can biological indicators be claimed to be a direct measure of pollution in assessing to what extent the amenities of a river have been affected by a discharge. Nevertheless, it has been shown that many of the changes in the nature of the water which adversely affect the amenities of a river, such as toxicity, de-oxygenation, changes in acidity, salinity, organic and mineral-salt concentrations, and in such physical factors as temperature, turbidity, and suspended-solid content, are all associated with observable changes in the stream-bed biota. Furthermore, the degree of change of the composition of the stream-bed community reflects the extent to which the properties of the water have been changed. Some changes in the biota may occur, reflecting changes in the nature of the water, which in themselves do not affect the amenities of the river and therefore do not constitute pollution. Such changes, however, would indicate a deterioration in conditions which, if continued, could result in pollution. Such indicators of potential or incipient pollution are surely of value in pollution-prevention work.

Apart from stream organisms being the indicators of pollution on the basis outlined above, in many cases organisms are the agents of pollution, their presence or absence, or change in abundance, directly affecting the amenities of the river. The discharge of pathogenic viruses, bacteria, protozoa, and worm eggs is probably the most common type of "biological" pollution. The development of undesirable growths, both in the water, in the form of water blooms or of micro-organisms imparting taste or odor, and on the stream bed, as sewage fungus or blanket weed, is also a common form of this type of pollution. The increase in abundance of certain invertebrate animals, such as some insects and snails which are vectors of diseases of man and his domestic animals, is another form of pollution where the agent is biological.

The protection of fishery interests has done much to safeguard the conditions of streams for other interests. The effects of discharges on fish, fish food, and their breeding grounds is again a biological consideration.

The several types of "biological" pollution can only be directly assessed by biological methods.

Pollution, because of its complex and manifold aspects and because of the many different interests involved, cannot be assessed by a few simple criteria, chemical, physical, or biological. The established value of stream-bed communities as indicators of changes in many factors affecting stream conditions involved in pollution, and the direct role of organisms as agents in other aspects of pollution, make the inclusion of biological observations essential in any comprehensive assessment of the pollutional condition of a river.

In considering biological assessment in relation to thermal pollution, it could be argued that changes in water temperature are more accurately and more readily measured by using a thermometer than indicated by changes in the biota. However, the indirect effects of increased temperature, more difficult to measure—such as reduction in oxygen or increase in toxicity—are readily indicated by changes in the biota. Furthermore, the degree of imbalance caused by the increased temperature, both directly and indirectly, is indicated only by changes in the biota. Having thus justified the whole aquatic biota as being significant, I do not intend to deal with the effects of increased temperatures on all members of the biota. Certain groups of organisms are favored as indicators, and most of these groups have already been justified for consideration on other accounts.

FISHES

Possible Effects of Raised Temperatures to be Considered

Probably the most obvious effect on fish of raised water temperatures is the direct autecological one affecting the distribution of each species according to their temperature-tolerance ranges. Under some circumstances, species normally restricted by low temperatures could extend their geographic range into artificially heated waters. It has been suggested that the American game fish bluegill *(Lepomis macrochirus)* and the large-mouth bass *(Micropterus salmoides)* be introduced into heated waters in Britain. Ross (1959) issues a word of caution on such ventures and quotes the case in which *Lepomis gibbosus* was introduced into France in mistake for the bluegill. *(Lepomis macrochirus).* The former fish (pumpkinseed) is a similar species to the bluegill but does not attain a size of interest to anglers. Of more concern, however, is the restriction of the distribution of a species in a river system by raised temperatures. At the lower temperature ranges, the cold stenothermal species are first to be affected; but with increasing temperatures, species will be successively eliminated according to their temperature tolerance. The vast amount of data, mostly from North America, on the temperature tolerance of fish will be considered, in relation to this problem, in the next section. Indirect effects of raised temperatures affecting other factors, such as oxygen and toxicity, need also to be considered.

Fish are nektonic and are not therefore confined to particular localities in the river, as are most stream organisms. The effects of temperature on migration, site selection, and the sensitivity of fish to temperature changes are therefore important. The effect on feeding, food organisms, over-all productivity, and disease will also be briefly considered.

Data Relevant in Predicting the Effects of Increased Temperatures on Fish

Relevant data is available from two main sources, field observations and laboratory experiments. Both have their disadvantages, which limit the usefulness of the data. Although field results on the range over which the species is found in nature provide more directly applicable data, there are limited reports of the higher extremes of temperatures in which the species survives and in which it thrives. The limitations of laboratory tolerance tests will be discussed in the relevant section.

Field Observations.—From different sources, including those reviewed by Wurtz and Renn (1964), Table 2 was compiled to show the temperature limits in which various fish are reported to be found. Data on the survival or death at different temperatures of short duration are also given.

Laboratory Experimental Evidence.—A vast amount of experimental data on the effect of temperature on fish has accumulated. Many of these investigations were carried out with different objectives and were not all ecologically oriented. This limits the value of this available data in determining the maximum permissible level of temperature for the survival of a specific fish population at an acceptable level.

There is also some confusion and consequent misunderstanding of the terms used in expressing the results of such work. The investigators using them undoubtedly understand what they mean by them; but nevertheless, such terms as "thermal death point," "thermal threshold," "upper lethal temperature," "maximum lethal temperature," "maximum tolerable temperature," "thermal index," "median tolerance limit," "24-hr. LD_{50}," "the ultimate lethal temperature," and "ultimate upper lethal temperature" may not be very meaningful to others or may be misleading, especially when it may not be appreciated how, in fact, such figures are obtained experimentally.

Apart from temperature requirements in relation to fish activity, spawning, and development, which will be discussed later, two types of temperature data are relevant to the problem of thermal discharges. Of ecological importance is the maximum temperature at which a species will survive for a prolonged period—indefinitely, were it otherwise immortal and did not age. This temperature we can consider as the *Maximum Tolerable Tem-*

TABLE 2. Temperature limits at which different fish reported to exist, survive and are killed in nature.

Above 40°C (104°F)

All fish in pond near Savannah, Georgia, killed when temperature reached 42°C (108°F)   (Tarzwell, 1956)

35°-40°C (95°-104°F)

The following **survived** in a shallow pool one afternoon when temperature rose to 38°C (100.4°F) and dropped to 35°C (95°F) the following day:

(Bailey, 1955)

      Micropterus salmoides (yearling) (largemouth bass)
      Lepomis cyanellus (green sunfish)
      Lepomis macrochirus (bluegill sunfish)
      Lepomis megalotus peltastes (longear sunfish)
      Lepomis gibbosus (pumpkinseed sunfish)

The following **were killed** by the same event:

      Amblopliles rupestris (rock bass)
      Umbra limi (mud minnow)
      Erimyzon sucetta (chub sucker)
      Notemigonus crysteleucas (golden shiner)
      Notropis heterodon (blackshin shiner)
      Notropis heterolepis (blacknose shiner)
      Notropis delicosus
      Pimephales notatus (bluntnose minnow)
      Ictalurus nebulosus (brown bullhead)
      Ictalurus notalis (yellow bullhead)
      Noturus miurus (brindled madtom)
      Noturus gyrinus (tadpole madtom)
      Etheostoma exile (Iowa darter)

The following have been reported active in waters at the temperature indicated.

(Trembley, 1960)

Fundulus diaphanus (Banded killifish)	38°C (100°F)
Lepomis macrochirus (Bluegill sunfish)	35°C (95°F)
Lepomis auritus (Yellowbelly sunfish)	35°C (95°F)
Lepomis gibbosus (Pumpkinseed sunfish)	35.5°C (96°F)

Etheostoma lepidum (green throat darter) taken from water
with max. 35°C (95°F)

(Hubbs and Strawn, 1957)

30°-34.9°C (86°-95°F)

Coarse fish population locally reduced in British rivers if mean daily temperature reaches 30°C (86°F)   (Alabaster, 1963)

30°C (86°F) temperature above which water in temperate zone streams should not be raised for prolonged periods.   (Cairns, 1956)

The following have been found active at the temperatures indicated:

(Trembley, 1960)

Carpoides cyprinus (Quillback carpsucker)	32°C (90°F)
Ictalurus nebulosus (Brown bullhead)	32°C (90°F)
Anguilla rostrata (American eel)	33°C (91°F)
Micropterus salmoides (Largemouth bass)	32°C (90°F)
Micropterus dolomieu (Smallmouth bass)	32°C (90°F)

Table 2—Continued

Poxomis nigromaculatus (Black Crappie)   34°C (93°F)
Etheostoma nigrum (Johnny darter)   31°C (88°F)
            Pair with maturing eggs.

25°-29.5°C (77°-86°F)
Most coarse fish of temperate zones found over this range.
    Ambloplites rupestris (Rock bass)
                usually found below 30°C (86°F)   (Trembley, 1960)
    Alosa pseudoharengus (Alewife)
                lived in water at 28.5°C (83°F)   (Trembley, 1960)
    Ictalurus punctatus (Channel catfish)
                active in water at 28.0°C (82°F)   (Trembley, 1960)

20°-24.9°C (68°-77°F)
Most coarse fish found throughout this zone, though below optimum for growth for many, e.g., **Cyprinus carpio** (Carp) and **Scardinius erythrophthalmus** (Rudd).
Spawning temperature for **Cyprinus carpio** 20°-22°C (68°-71.6°F)
                                                       (Varley, 1967)
    Micropterus dolomieu eggs incubate successfully at 25°C (75°F)
                                                       (Webster, 1945)
    Salvelinus fontinalis (Brook trout) although found at higher temperatures (24°C) (75°F) are not favored in competition with coarse fish at temperatures above 20°C (68°F). This is probably true of most salmonid fishes.   (Tarzwell, 1956)

15°-19.9°C (59°-68°F)
This is the zone favoring the salmonid fishes. Good trout streams should have temperatures not exceeding 68°F. Temperature range for the spawning of some fish:
    Rutilus rutilus (Roach)   15-16°C (59-61°F)
    Cyprinus carpio (Carp)   18°C   (64.4°F)
    Oncorhynchus tshawytscha (Chinook salmon)   15.5°-18.5°C (60°-65°F)
                                                       (Varley, 1967)

10°-14.9°C (50°-59°F)
Below the optimum temperature for growth for most fish except some salmonids. Below temperature for spawning for some fish. Optimum temperature for growth for **Salmo trutta** (Brown trout), **Salmo gairdneri** (Rainbow trout) and **Salvelinus fontinalis** (brook trout) 12°C (55°F).
Salmo salar (Atlantic salmon) do best at 16°-18°C (60°-65°F).
Fish reported from waters with temperature range 10°-15°C (50°-59°F).
                                                       (Ferguson, 1958;
                                                        Rawson, 1960)

        Coregonus clupeaformis (lake whitefish)
        Leucichthys sp.
        Esox lucius (northern pike)
        Stizosredion vitreum (walleye)
        Catostomus commersoni (common sucker)
        Catostomus catostomus (longnose sucker)
        Lota lota (burbot)

Table 2—Continued

Salvelinus namaycush (lake trout)
Oncorhynchus nerka (sockeye salmon)

Below 10°C (50°F)
   Salmonids which feed actively at 0°-1°C (32°-33°F)
      Salmo trutta (brown trout)
      Salmo gairdneri (rainbow trout)
      Salvelinus fontinalis (brook trout)
Fish requiring temperatures below 10°C (50°F) for spawning:
      Salmo trutta (brown trout)   4°-10°C (39°-50°F)
      Salvelinus fontinalis (brook trout)   3°-7°C (37°-44°F)
      Oncorhyndrus tshawytscha (chinook salmon)   5.5°-14.5°C (42°-58°F)
      Esox lucius (northern pike)
      Thymallus vulgaris (grayling)   about 10°C (50°F)
      Phoxinus phoxinus (minnow)
      Stizostedion vitreum (walleye)   7°-10°C (45°-50°F)
Fish found inhabiting waters having maximum temperature less than 10°C (50°F)
                                                            (Ferguson, 1958)
      Pomolobus pseudoharengus   4.4°-8.8°C
      Salvelinus namaycush (lake trout)
      Osmerus mordax (American smelt)
      Leuchichthys artedi

perature for the species. Such figures are important in predicting the effects of an increase in general temperature level of a water above normal by a thermal discharge. Experimentally, they may be determined as the highest temperature at which no test animal dies over a prolonged period. Such tests, however, are time-consuming and are therefore rarely carried out.

More experimentally convenient are tests in which the temperature is determined at which a definite proportion, usually half, of the test animals die within a definite period of time, e.g., 24 hours or 48 hours. This corresponds with the established procedure used in fish-toxicity tests, and the results are sometimes quoted in similar terms of the "median tolerance limit" (TLm), being the temperature at which half the fish survive for a stated period, e.g., "24-hour TLm." This figure is then sometimes taken as the threshold value but, as pointed out by Brown et al. (1967), in case of fish-toxicity work, this value, as determined, is above the realistic threshold, as generally understood in toxicology, for half of the test animals! They suggest the use of the term "Median Lethal Concentration (LC50)" to conform with earlier and standard toxicological practice. For the purpose of thermal studies, I suggest the term "Median Lethal Temperature (LT50)" as being equivalent. Such a term qualified by the duration of the test, e.g., "24-hour Median Lethal Temperature" (24-hour LT50) is quite unambiguous and not likely to be misinterpreted or misapplied.

Species, of course, have upper and lower thermal death points and consequently we should recognize the terms "Lower Median Lethal Temperature" and "Upper Median Lethal Temperature." Furthermore, since, as discussed previously, the LT50 value is dependent upon the acclimation temperature up to (or down to) a certain temperature, we can refer to the "ultimate upper (or lower) LT50," as indicating the value above (or below) which further acclimation has no effect. Finally, an "incipient LT50" is a value which could be increased (or decreased) by acclimation.

We now seem to have derived more terms than those under criticism, but at least we have defined them!

The LT50 is of value in comparative work and in experimental studies investigating the effects of other factors on the effect of temperature. It is not, however, the temperature at which the species could be expected to survive indefinitely: i.e., the maximum tolerable temperature. It is, however, of applied significance in the case of short-term high-temperature conditions which occur in practice. In spite of the difficulties outlined above, an attempt is made in Table 3 to summarize the data from various sources. Since it is not always clear exactly what value is quoted, the term "Thermal Death Point" is used in the table. The figures are listed under different acclimation temperatures when these are quoted.

Further difficulties arise in the use of such data in predicting the effects of increased temperatures when it is appreciated that the fish not only must survive but, to provide an amenity or economic fishery, they must grow and multiply. To survive, an organism requires a basic metabolic activity, or energy consumption; when active, additional metabolic activity and energy consumption is required. The energy requirement for basic or resting metabolism increases with increasing temperature, thus less is available for activity. This introduces the concept of "scope for activity," which implies the desirability of maintaining the temperature well within the tolerance range. At optimum temperature, the species has full scope for activity. Where it is not possible to provide this, half or three-quarter scope for activity could be acceptable. The brook trout *(Salvelinus fontinalis),* with a temperature range of 0° to 25.3° C. (32° to 77.5° F.) was found by Baldwin (1956) to consume most food and grow more at 13° C. (55.4° F.). Gibson and Fry (1954) found that the lake trout *(Salvelinus namaycush),* having an upper lethal temperature of 23.5° C. (74° F.), displayed their maximum sustained swimming speed at 16° C. (60.8° F.).

TABLE 3. Upper Thermal Death Points and Temperature Preferenda of Some Fishes
(See remarks in text regarding limitations)

Species		Approximate Acclimation Temperatures								References
		Not known	5°C	10°C	15°C	20°C	25°C	30°C	Ultimate	
Salvelinus namaycush (Lake Trout)	Lethal								23.5	Gibson and Fry, 1954
	Pref.								12	Ferguson, 1958
Oncorhynchus gorbuscha (Pink Salmon) (fry)	Lethal		21.3	22.5		23.9				Brett, 1956
	Pref.	11.7								Ferguson, 1958
Salmo gairdneri (Kamloops) (Rainbow Trout)	Lethal			24						Black, 1953
	Pref.	13.5	16	15	13	11				Garside and Tait, 1958
		13.0								Ferguson, 1958
Salvelinus fontinalis (Speckle Trout, Brook Trout or American Char)	Lethal			24					25.2	Gibson and Fry, 1954; Black, 1953
	Pref.		12 (8 in winter months)							Sullivan and Fisher 1953
	Pref.	14-16								Ferguson, 1958
Oncorhynchus tshawytscha (Chinook Salmon) (fry)	Lethal				25	25.1				Brett, 1956
	Pref.	11.6								Ferguson, 1958
Oncorhynchus nerka (Sockeye Salmon) (fry)	Lethal	22	22.2	23.4		24.8				Brett, 1956; Black, 1953
	Pref.	14.5								Ferguson, 1958
Salmo trutta (alevins)	Lethal		25.5							Spaas, 1961

Table 3—**Continued**

Species										References
(Brown Trout)	(fry)		22.5	23						Bishai, 1960
	(yearling)		26							Spaas, 1961
	(parr)		29							Spaas, 1961
		Pref.	12.4-17.6							Ferguson, 1958
Oncorhynchus keta (Chum Salmon)	(fry)	Lethal		23.1	23.7					Brett, 1956
		Pref.	14.1							Ferguson, 1958
Salmo salar (Atlantic Salmon)	(alevin)	Lethal	27.5							Spaas, 1961
	(yearling)		28.5							Spaas, 1961
	(parr)		29.8							Spaas, 1961
	(parr)		32.5-33.8							Huntsman, 1942
									27	Gibson and Fry, 1954
										Black, 1953
Perca flavescens (Yellow Perch)		Lethal	26.5	29 (at other times)						Hart, 1947
			(winter) 21.2	27.7						Wurtz and Renn, 1964
			33 (Acc-8)	29.8						Fry, 1957
		Pref.	18.5	20.6	23	23	24.5	25.4		Ferguson, 1958
Rutilus rutilus (Roach)		Lethal	33.5			29.5	30.5	31.5		Cocking, 1959
		Pref.	23-24							Alabaster and Downing, 1958
Tinca tinca (Tench)		Lethal	29-30							Weatherley, 1959
		Pref.	35.2							Varley, 1967

Table 3—Continued

Species										References
Micropterus salmonids (Largemouth Bass)	Lethal	28.3		35			36.1-36.7	36.4		Black, 1953; Trembley, 1960; Brett, 1956
	Pref.	26	30-32			32.5				Fowler; Ferguson, 1958
Perca fluviatilis (Perch)	Lethal	30-31	32.8							Weatherley, 1963; Varley, 1967
Cyprinus carpio (Carp) (small)	Lethal	38-39				31-34	35.8			Black, 1953; Meuwis and Heuts, 1957
(large)		35-36								Meuwis and Heuts, 1957
	Pref.			17	25	27	31	32		Pitt, Garside, and Hepburn, 1956
Lepomis gibbosus (Pumpkinseed)	Lethal		32.8			39				Trembley, 1960
	Pref.								31.5	Ferguson, 1958
Lepomis macrochirus (Bluegill)	Lethal		31.7	35			36.1-37.2		39.5	Trembley, 1960
	Pref.								32.2	Ferguson, 1958
Carassius auratus (Goldfish)	Lethal	30.8				34.8	36.6		38.6	Black, 1953
	Pref.								41	Fry, Brett, and Clawson, 1942
									28.1	Ferguson, 1958
Tilapia mossambica	Lethal								38.2	Allanson and Noble, 1964
									41.9	Long, et al., 1963

It is possible that, although some fish may feed more actively at temperature approaching the maximum of their range, their growth rate is less than at lower temperatures because of the demands of the higher basic metabolism.

Different stages of development of the same species often exhibit different LT50 values; the developing eggs usually have a lower upper-LT50 value than the adults. Figure 3 illustrates the difference in tolerance ranges of different stages of selected species. The young also are often more sensitive to high temperatures but in some cases are more resistant than the adults.

Factors Influencing Temperature Tolerance of Fish

As mentioned in the introductory section, in nature, there is an interaction of factors. Craigie (1963) found that the thermal resistance of yearling trout *(Salmo gairdnerii)* reared in soft water was greater than those reared in hard water when tested in soft and saline waters. Increased salinity of the test water resulted in an increase in thermal resistance. Hoar (1956) reported that goldfish *(Carassius auratus)* which had been conditioned to 8 hours of light per day died more rapidly when subjected to a lethal temperature of 35.6° C. (95.9° F.) than did fish conditioned to 16 hours of light per day. This effect of photoperiodism may be relevant when considering the lethal effects at different seasons of the year.

Indirect Effects of Temperature Affecting other Factors

In many cases, heated water is discharged to already polluted rivers, and the indirect effect of temperature under such conditions may be more important than the direct effect of increased temperature.

Oxygen Concentration.—At increased temperatures, the oxygen requirement of fish increases. Generally, fish with a high oxygen requirement are cold-water forms, those with low requirement are eurythermal (Varley, 1967). Thus, at the same low oxygen concentration in a polluted river, the effect is likely to be greater at higher temperatures. Downing and Merkens (1957) found that rises in temperature within the range 10° to 20° C. (50° to 68° F.) reduced the resistance to lack of oxygen of *Perca fluviatilis* (perch), *Rutilus rutilus* (roach), *Cyprinus carpio* (domesticated mirror carp), *Tinca tinca* (tench), *Leuciscus leuciscus* (dace). In the case of *Salmo gairdnerii* (rainbow trout), although there was a marked lowering of resistance between 10° C. (50° F.) and 16° C. (60.8° F.), there was no significant difference between 16° C. (60.8° F.) and 20° C. (68° F.).

The effect of increased temperatures on the oxygen level in polluted and non-polluted waters will be mentioned in the section on the effects of temperature on the chemistry of water.

Toxicity.—It is generally thought that the toxicity of pollutants increases with increasing temperature. Although this may be so for many substances, there are important exceptions to this. Alexander et al. (1935) found that the toxicity of cyanide solutions to salmon smolt *(Salmo salar)* and rainbow trout *(Salmo gairdnerii)* increased rapidly with rise in temperature over the range of 6° to 19° C. (42.8° to 66.2° F.). It has been established that the survival time of many fish in toxic solutions decreases with increase in temperature, there often being a linear relationship between the log. of the median period of survival and temperature. Wuhrmann and Woker (1953) found this to be so for minnow *(Phoxinus phoxinus)* in toxic concentrations of ammonia but with chub *(Squalius cephalus)*, although the survival time decreased with a rise in temperature from 10° C. (50° F.) to 15° C. (59° F.), there was no further decrease in survival time with a further increase in temperature to 25° C. (77.6° F.). Lloyd and Herbert (1962), however, considered that, in the case of poisons such as zinc, which have a threshold of toxic concentration, this value did not appear to be affected by temperature. They also stated that the threshold concentration for un-ionized ammonia and phenols are little, if at all, affected by temperature changes within the physiologically tolerable range, although in toxic concentrations of these poisons (above the threshold value), fish survive longer at lower temperatures. They concluded that, on the whole, the effect of temperature on the threshold concentration of a poison would be small compared with the effect of other environmental factors. Later work by Brown et al (1967) showed that, although at higher concentration of phenols there was a decrease in survival time of rainbow trout with increase in temperature, at lower concentrations, the toxicity, as measured by the 48-hour LC50, increased as temperature decreased.

Responses of Fish to Temperature Differences

Unlike the benthic invertebrates, which are more or less confined to a given locality of the river bed and are therefore subjected to changing quality of water flowing over them, fish are able to respond to changing conditions and have powers of selecting their locality. It is probable that, in many cases, thermal discharges, by increasing the water temperature above the optimum, would affect local fish populations more by affecting the site selection of fish than by fish kill. Also, in the case of migratory anadromous fish, the effect of temperature could be significant in affecting the selection of course in the upstream migration to the spawning grounds.

Temperature Sensitivity.—Many fish in nature have been observed to select a certain temperature and therefore have the powers of "temperature selection." The natural distribution of the brook trout *(Salvelinus fontinalis),* for example, has been found to be restricted to streams where the maximum temperature does not exceed 19° C. for any considerable period, although experimentally it can withstand higher temperatures. Near this maximum temperature, a slight rise in temperature, such as might result from forest clearing, could change a stream favorable to trout into one unfavorable. Such critical conditions are reported to exist in natural trout streams in Michigan and in the Smoky Mountain region of Tennessee (Creaser, 1930). There is evidence that fish possess cutaneous thermal receptors which are not, however, in the trunk lateral line. The cerebellum is indicated as being concerned with temperature reception (Sullivan, 1954). The frequency of movement of a fish moving in a temperature gradient is least in the selected region. Moderately rapid changes of temperature do not elicit locomotor responses in resting fish until very high temperatures are reached, but they do affect the frequency of movements of active fish. It would appear, therefore, that fish could be "caught napping" by a shock increase in temperature.

Temperature Preferenda.—The thermal preference or "temperature preferenda" has been determined for many fish by presenting them with a temperature gradient under laboratory conditions. Ferguson (1958) found that with yellow perch *(Perca flavescens)* the temperature selected was influenced by the acclimation temperature. In comparing the laboratory-determined temperature preferenda with field observations, the same worker found good agreement with fish having lower final preferenda, *Salvelinus fontinalis, S. nanaycush, Salvelinus hybrid* and *Coregonus clupeaformis;* but fish having higher final preferenda, such as *Micropterus salmoides, M. dolomiea,* and *Lota lota lacustris,* had higher values in the laboratory than were exhibited in natural conditions. The differences were ascribed to the use in laboratory tests of the young fish, which probably had higher preferenda than the older fish. This was found to be so with *Perca flavescens.* The temperature preferenda for several fishes from data quoted by Ferguson and elsewhere is included in Table 3.

Reaction of Migratory Fish.—Collins (1952) investigated the effect of temperature on the orientation of migrating andromous fish *Pomolobus pseudoharengus* (alewife) and *P. aesticalis* (glut herring) during the upstream migration through the Herring River to the spawning grounds. In the test, the migrating fish were presented with a choice of two channels carrying water of different temperatures. It was found that when the temperature difference continuously exceeded 0.5° C., 77 percent of the fish selected the warmer water over a range of 11.1° C. (52° F.) to 23.3° C. (74° F.). Applying these results to the thermal-pollution problem, this response could lead them into hotter waters upstream.

The Effects of Heated Water Discharges on Fish

Direct evidence of the effects of thermal discharges is being obtained by surveys of affected rivers and lakes. In some cases, there are complicating factors to be taken into account. For example, heated-water discharges from steel mills may contain toxic substances, such as cyanide. Even discharges from electricity-generating stations may contain substances not present in the intake water. The use of chlorine to remove slimes from condenser tubes may cause the effluent to be toxic at times (Trembley, 1960). Alabaster (1963) has reported the results of surveys carried out below several electricity-generating stations in Britain ranging from 40-mw to 600-mw capacity. The sites chosen were on rivers where the fish population was not otherwise suppressed by pollution. The temperature conditions at the different stations presented in the paper may be summarized in tabular form to show the temperature ranges involved:

TABLE 4. Summer Temperatures of Normal River Water and Heated Water Discharges at Six Electricity-Generating Stations in Britain

Station	Mean Temperature ° C. (° F.)			
	Normal River Water		Effluent	Difference
A	20.2	(68.3)	26.5	(79.7)	6.3
B	16.9	(62.5)	27.4	(81.4)	7.5
C	18.4	(65)	27.0	(80.6)	8.6
D	18.4	(65)	28.8	(83.8)	10.4
E	20.4	(68.5)	27.2	(81)	6.8
F	21.5	(70.7)	30.0	(86)	8.5

(From Alabaster, 1963)

The temperatures of the rivers downstream depended upon the degree of mixing. In all cases, a restricted zone consisted of undiluted heated effluent, and in cases where only a fraction of the flow was used, the hottest water persisted downstream as a surface layer, leaving the stream bed little affected. Where most of the river flow was used, the temperature conditions downstream were uniform (Stations B and F).

Under summer conditions, tests were carried out in which different fish were transferred from normal river temperatures to cages in the heated effluents. In these tests, all trout and a substantial proportion of roach *(Rutilus rutilus)*, perch *(Perca fluviatilis)* and gudgeon *(Gobio gobio)*, but none of the tench *(Tinca tinca)*, and carp *(Cyprinus carpio)* were killed within 24 hours.

During the investigations, a fish-kill only occurred once under normal operating conditions. At Station F, the effluent temperature rose from 30.5° C. (86.9° F.) at 1200 hours to 36.5° C. (97.7° F.) at 1500 hours. Dead fish, gudgeon and roach, were observed floating downstream at 1630 hours. It is recorded that most of these were smaller than those known to be present in the locality, suggesting that the larger fish near the outfall escaped the lethal temperature but the smaller fish could not.

It was concluded that, under the conditions then prevailing in Britain, although lethal conditions may exist in the immediate vicinity of discharges in the summer months and result in occasional fish death, especially of smaller fish, the chances of fish kills were rare. The larger fish would avoid such waters and, in cases where thermal stratification existed, coarse fish were found in great abundance and variety under the heated layer. Where the river temperature as a whole is increased, it was concluded that fish populations may be locally reduced when the mean daily temperature reaches 30° C. (86° F.) and increased when the temperature is increased to not more than 26° C. (78.8° F.). In predicting the lethal effects from laboratory data, it was found that closer correlation was obtained when maximum river temperatures were plotted than when mean figures were used.

Trembley (1960) studied the effects of heated-water discharges into an otherwise unpolluted stretch of the Delaware River, which resulted in summer maximum temperatures exceeding 38° C. (100° F.) in the river 1500 feet downstream of the discharge. He found that most species were eliminated from the zone of maximum temperature during the warmer months. During the colder months, however, fish collected in the heated water and there was an extended period of feeding activity of many species. There was, over-all, a change in the proportional species composition of the fish fauna, some species decreasing in abundance and others increasing. All normal inhabitants of the Delaware River were found living in the heated zone at one time of the year at least, with the exception of the American shad *(Alosa sapidissima)*. Since young were found to be present upstream of the heated zone, adults of this anadromous species must be capable of migrating through the heated zone. This also applied to other anadromous species, *Petromyzon marinus* (sea lamprey), and *Roccus saxatilis* (striped bass) and to the catacromous American eels *(Anguilla rostrata)*.

Both these accounts refer to relatively unpolluted rivers, the effects of heated-water discharges into organically polluted rivers, subjected to oxygen depletion may be quite different.

MACRO-INVERTEBRATES

Temperature Tolerance

Probably temperature affects the distribution of macro-invertebrates in a similar way to that responsible in the distribution of fish, although less information is available on their tolerance limits. Data on the occurrence of species in hot springs, already discussed, gives some indication of groups tolerating high temperatures, e.g., Protozoa, Rotifera, Nematoda, Annelida, and Ostracoda.

Boycott (1936) considered that, although mollusca are favored by moderately warm calcareous waters, a continuous temperature of 30° C. (86° F.) could not be tolerated by British species. Some groups, such as Plecoptera (stoneflies), exhibit marked cold stenothermal characters.

Sprague (1963) investigated the temperature tolerance of four freshwater crustaceans, *Asellus intermedius, Hyalella asteca, Gammarus fasciatus* and *G. pseudolimnaeus*. He found that raising the acclimation temperature raised the 24-hour LT50 value in case of all but *H. asteca*. The ultimate values being *A. intermedius* and *G. fasciatus,* 34.6° C. (94° F.); *H. asteca,* 33.2° C. (91.6° F.); and *G. pseudolimnaeus,* 29.6° C. (85.2° F.). Considering the different habitats of these crustaceans, these values are not as different as might be expected. Furthermore, their thermal tolerance appeared greater than needed in relation to the distribution in nature. This suggests that temperature is not the major factor influencing distribution; oxygen was considered to be more important. Walshe (1948) investigated the thermal tolerance of seven midge larvae. She found that the species associated with lentic conditions and therefore subjected to higher temperatures, had higher thermal resistance than those from streams. The stream species with the highest thermal resistance *(Chironomus alvi-*

manus) was also found in stagnant waters. Thus, there appears to be a closer correlation between temperature tolerance and habitat in this group.

Insects with aquatic larval stages have their life cycle timed to seasonal changes, and the artificially increased temperatures may thus upset the cycle by causing the premature emergence of adults when aerial temperatures and other conditions are unsuitable.

Chironomid larvae increase their growth rate with increasing temperature over a wide range. Thus, increased temperatures throughout the winter increase the supply of fish food. Thus, fish capable of living in heated waters are not likely to be short of food because of the higher temperatures.

Surveys below Heated Discharges

Most macro-invertebrates in rivers are benthic and may not therefore be subjected to the full effects of the heated water in all cases. Mann (1965) investigated the effect on the macrobenthos of the River Thames at Reading of the discharge of heated water from Earley electricity-generating station. The benthos was subjected to normal river temperature for part of the night and to temperatures up to 12° C. above normal for varying periods of the 24-hour period. He found no evidence of animals having been killed by these conditions. He reported an increase in the mollusca population but a decrease in crustacea, leeches, and midge larvae. From his observations, he considered that the lethal temperature for many benthic animals is between 32° to 35° C. (89.6° to 95° F.). He discovered the presence of an exotic worm *Branchiura sowerbyi* in the heated waters, a native of southeast Asia. The snail *Physa acuta,* another tropical species, has appeared in heated water in temperate countries, such as Britain and Poland.

Trembley (1960) found that, below the thermal discharge into the Delaware River, there was a reduction in invertebrates in the zone of maximum temperature during the summer. There was, however, a recolonization of this zone during the colder months by drift fauna from upstream. A considerably higher standing crop of macro-invertebrates existed downstream from the zone of maximum temperature than in the unheated section of the river. Chironomids appeared to be the most tolerant group. There was some evidence of earlier emergence of insects in the heated water zone.

The effect of the shock temperature increases on organisms in the condenser-cooling water has been investigated by Markowski (1959). Amongst the organisms which apparently survived this treatment were the invertebrates Turbellaria (flatworms), Nemotoda (5 spp.) Rotifera (5 spp.),

Tubifex, Crustacea (16 spp.), Diptera larvae including *Tanytarsus, Tendipes,* and *Simulium,* and young snails, *Lymnaeidae.* No detrimental effect was noted except some mechanical damage to a small percentage of the larger crustacea, such as shrimps.

Wurtz (1961) developed a method, using benthic macro-invertebrates as indicators, to assess the degree of imbalance caused by thermal discharges into the Schuylkill River.

MICRO-ORGANISMS

Although the effect of raised temperatures on aquatic micro-organisms has generally received less attention than the effect on fish, changes in microbial populations resulting from increased water temperatures may well prove to be more important in affecting water quality and the general productivity of aquatic ecosystems.

As illustrated in an earlier section, the processes by which temperature affects microbial populations is complex, and the effect of raised temperatures is not likely to be predicted in simple autecological terms of temperature tolerance. It is convenient to consider the micro-organisms under two headings, the Phototrophes (algae) and the Heterotrophes (bacteria and fungi), protozoa being represented in both.

Algae

As discussed earlier, the composition and succession of algal populations are probably the result of the joint effects of temperature, light intensity, and nutrition. Temperature is probably important in deciding the outcome of interspecific competition and therefore the temperature optima of species is more significant than tolerance ranges. Within the algae, the diatoms are reported to grow best at 15° to 25° C. (59° to 75.2° F.); chlorophyceae— the green algae—at 25° to 35° C. (75.2° to 95° F.); and cyanophyceae— the blue-greens—at 30° to 40° C. (95° to 104° F.); but there are wide specific differences within these groups and some notable exceptions, the troublesome blue-green alga *Oscillatoria rubscens* being recognized as a cold-water form. Work reported by Cairns (1955) showed that some diatoms grew best over a narrow temperature range (4° C.), e.g., *Nitzchia filiformis,* whilst others had a wide range, e.g., *Gomphonema parvulum,* with an optimum at 22° C. (71.6° F.) but still good at 34° C. (93.2° F.).

Attached Algae (Benthos and Periphyton).—These algal communities are the most significant ones in rivers, whereas the plankton are more important in lakes. Trembley (1960) has studied in some detail the effects of increased temperatures on periphyton communities in the Delaware River. For this purpose, he designed his "Pralgometer," a modification of Dr. Ruth Patrick's "Diatometer". He found that the species composition of periphyton communities was considerably changed by the increased temperatures. The sensitivity of periphyton communities as a whole suggests that it could be used in indicating the degree of departure from the normal biological conditions induced by thermal discharge.

All four groups of blue-green algae and diatoms of the family Fragillariaceae increased in abundance in the heated water, but other algae are reduced. The over-all algal productivity was probably greater in the heated water than in the normal river but was in the form of the less desirable blue-green algae. They were, however, probably suppressed by the discharge of chlorinated waters. A similar investigation of the effects of heated-water discharges in the Vistula in Poland is at present in progress and should provide useful data, judged from unpublished preliminary results which I have been privileged to examine. Previously reported results from Poland, in which Stangenberg and Pawlaczyk (1960) investigated the zone of an organically polluted river (Nysa Luzycka) heated by the discharge from an electricity-generating station so that the seasonal temperature range was 5.5° C. (41.9° F.) to 36° C. (96.8° F.), showed a marked decrease in periphyton in the summer period when the temperature exceeded 30° C. (86° F.). Only a few species were more abundant during the period of high temperature, *Navicula cuspidata, N. ambigua, Tabellaria spp.* and the filamentous alga, *Ulothrix subtilis;* these formed an intense growth. Markowski (1960) observed good growth of algae *(Enteromorpha)* near the outlet of an electricity-generating station but none near the inlet.

Phytoplankton.—Appourchaux (1952) investigated the effect of discharging treated waters on the flora and fauna of the River Seine. The temperature increase was approximately 7° C. He concluded that the heated water did not bring about a spectacular modification of the plankton community. Although the ecological conditions may be ecologically modified, reducing the numbers of some species, there was no over-all suppression of the phytoplankton population. Up to a maximum of 30° C. (86° F.), both zooplankton and phytoplankton remained sufficiently abundant. In the winter, the heated water favored the development of certain species.

Poltroacka-Sosnowska (1967) compared the phytoplankton in three Polish lakes having different temperature ranges. Lichen Lake received thermal discharges from the electricity-generating stations at Konin and had a temperature range of 7.4° C. (45.4° F.) to 27.5° C. (81.5° F.). Slesin Lake was not influenced by thermal discharges and had a temperature range of 0.8° C. (33.5° F.) to 20.7° C. (69° F.). The third lake was only slightly influenced by thermal water. It was found that Lichen Lake, the warmest lake, supported the richest phytoplankton flora: 285 forms; and Slesin Lake, the least number: 198. In contrast with the other lakes, the phytoplankton flora of Lake Lichen was comparatively constant. It was observed that, as the temperature of Lichen Lake rose, the numbers of phytoplankton species increased. The characteristic dominant forms in Lichen Lake were the diatom *Melosira granulata* and the blue-green alga, *Microcystis aeroginosa*. In the cold water of Lake Slesin, the diatom *Stephanodiscus astraea* was the dominant form.

Patalas (1967) compared the productivity of Lichen Lake with that of a natural cold-water lake in the same lake system. It was found that the primary productivity of the heated lake (7.3 g/m²/d) was almost twice that of the cold lake, 3.75 g/m²/d. Secondary productivity in the form of phytophagous crustacea and rotifers was 4.5 g/m²/d in the heated lake, compared with 1.06 g/m²/d in the unheated lake. Under these circumstances, therefore, thermal "pollution" could be a valuable asset.

Bacteria and Fungi

There appears to be the least available data on the effects of heated discharges on this important group of organisms. The results of pure culture work on these micro-organisms is rarely applicable to field conditions.

Appourchaux (1952) found that the heated discharge into the Seine which increased the temperature by 7° C. to a summer maximum of 28° C. (82.4° F.) had no definite appreciable effect on bacterial groups they examined. These were: The saprophytic forms growing at 22° C. after 72 hours' incubation; enteric organisms developing at 37° C. in 24 hours; *E. coli* by filter-membrane technique; and the faecal Streptococci. These did not include the psychrophilic forms, which are of ecological significance as opposed to the public-health significance of the groups tested. There was also very little effect on the anaerobic sulphur-reducing and the non-sulphur-reducing bacteria.

Smyk and his co-workers (1959) have carried out detailed investigations on the effects of heated discharges on the microbiology of rivers. They reported that the heated water having temperatures 7° to 15° C. above that of the natural river water destroyed the biochemical activity of the local psychrophilic microflora. The numbers of psychrophilic heterotrophic microflora responsible for the breakdown of organic matter in the polluted river were reduced in the heated water. These included the proteolytic and some cellulose-decomposing bacteria. Stangenberg and Pawlaczyk reported similar effects (1960). There was, however, a slight increase in the mesophilic forms of these groups of bacteria. In the case of the Vistula, one of the rivers concerned, these results may be the effect of the increased temperatures on toxicity. There may also be a direct effect of temperature, as suggested by the workers.

These investigations are being continued and, if the above observations are confirmed, they should be significant in relation to assessing the effects of heated discharges on the waste-assimilative capacity of rivers.

Slime Infestations.—It is often considered that thermal discharges will increase the growth of sewage fungus *(Sphaerotilus)* in polluted rivers (Laberge, 1959). Although under some winter conditions this could well be so, the general position is more complicated. In non-heated waters, sewage-fungus growths are commonly associated with winter conditions and low temperatures. Optimum ranges have been variously quoted as 7° to 17° C. (44.6° to 62.6° F.) (Hohn, 1956) and 25° to 37° C. (77° to 98.6° F.) (Harrison and Heukelekian, 1958). Because of the higher rates of oxidation at higher temperatures, the stretch of stream affected is shortened at increased temperatures. However, even in the shorter stretch affected, the growths are less profuse than in colder weather, presumably due to the higher rates of basic metabolism utilizing a higher proportion of the available nutrient material. The maximum standing crop of micro-organisms is usually obtained at temperatures lower than those producing maximum growth rate (Lamanna and Mallette, 1953).

Reduction in the amount of *Sphaerotilus* in the heated zones of rivers has in fact been recorded by the Polish workers (Stangenberg and Pawlaczyk, 1960; Smyk and Cienciala, 1959).

WATER CHEMISTRY

It is best to consider the effects of increased temperatures on water chemistry under two headings: natural unpolluted waters and organically polluted water.

Effects on Natural, Unpolluted Waters

Theoretically, the only chemical changes of ecological significance likely to occur in natural waters by increased temperatures is a reduction in the dissolved oxygen due to the lower degree of solubility. This could only result in loss of oxygen in waters at saturation level. In fact, a rise in temperature under these circumstances would probably result in supersaturated conditions for a period, with little loss of oxygen. Furthermore, even if reduction in dissolved oxygen did occur as determined by the change in solubility, it is doubtful whether the reduced oxygen concentration would affect the organisms capable of withstanding the associated temperature. At 30° C. (86° F.) there is still present 7.53 p.p.m. oxygen, whereas at 15° C. (59° F.) there would be 9.76 p.p.m. at saturation. Critical temperatures are likely to occur in daytime during the summer when the photosynthetic activity of algae and plants is likely to enhance the oxygen supply.

It is commonly found that, with non-polluted waters, the heated-water discharges from electricity-generating plants has the same concentration of dissolved oxygen as the feed water; and in some cases, due to aeration in the cooling towers, it may be slightly increased. In such cases, there should be no reduction in oxygen downstream of the discharge.

Effect on Organically Polluted Waters

The indirect effects of increased temperatures on the oxygen balance in organically polluted streams is likely to be one of the most serious aspects of thermal pollution, in most cases. As the water temperature increases, the rate of biological oxidation increases proportionately, thus increasing the uptake of oxygen from that available in the water. This results in a greater oxygen depletion; but because of the higher rate of purification, a shorter length of river is affected. This is so only if complete de-oxygenation does not occur when the aerobic processes of purification break down. However, well before this condition is reached, the oxygen level in the stream may be sufficiently low to be harmful to the biota.

It is generally assumed that this effect of temperature on biological oxidation is operative over a wide range of temperatures; but Gottas (1948) found that the temperature coefficient of the reaction decreased as temperature increased, the rate of reaction reaching a maximum at 30° C. (86° F.). If the results of the adverse effects of heated waters on the microbial populations in organically polluted rivers are confirmed, then they could be significant in this context.

The rate of re-aeration is also affected by temperature and also by the oxygen deficit in the water. The rate of solution of oxygen increases linearly with temperature over the range of 0° to 35° C. (32° to 95° F.). Truesdale and Vandyke (1958) concluded that a mean value for the temperature coefficient of the exchange coefficient in flowing waters was 1.5 percent per ° C. The interaction of these effects of temperature on the oxygen balance have been discussed by Laberge (1959) and by Hoak (1952).

A further consideration of the effects of increased temperatures on organically polluted streams is that of the anaerobic microbial activity in the bottom sludges. This is likely to be increased, resulting in the release of undesirable substances into the water and the rising of sludge during the hotter months.

Davies (1966) has studied in some detail the chemical changes occurring in cooling towers at six British Central Electricity Generating Boards' stations. The average degree of cooling in the towers was between 7° to 9° C. Because of loss of water by evaporation, there was some concentration of dissolved salts, such as sulphates and chlorides. He found the concentration factor involved was usually between 1.15 and 1.3. Dissolved carbon dioxide was scrubbed out of the water, resulting in changes in hardness and pH values. There was a slight increase in pH; this was thought to be due to the loss of the loosely-bound CO_2 from the bicarbonate, moving the equilibrium of the equation:

$$Ca(HCO_3)_2 \rightleftharpoons CaCO_3 + CO_2 + H_2O$$

to the right.

This would also account for the change in hardness.

Nitrification of ammonia occurred, decreasing the ammoniacal nitrogen value and increasing the nitrate concentration. This would tend to decrease the pH. There was no evidence of loss of over-all nitrogen as gaseous nitrogen to the atmosphere. Aeration in the towers tended to increase the dissolved oxygen. With organically polluted waters, biological oxidation of the organic matter took place. There was limited concentration of suspended solids; but with polluted water, biosynthesis resulted in increased solids, which could, of course, be removed by settlement.

In Conclusion

The foregoing review of the ecological significance of temperature as a factor influencing the distribution, abundance, and activity of organisms naturally and the effects of elevated temperatures induced by man's industrial activities makes evident the complexity of the problem of thermal discharges. For good or bad, increased temperatures will affect aquatic biocoenoses. Whether the effects are adverse or beneficial to man will depend upon the temperature range involved in relation to the use the water body best serves man's interest, remembering man's interests are best served by preserving natural ecosystems as far as is compatible with his other requirements. In practice, this involves deciding by what use a water body can best serve a community aesthetically and recreationally and economically, and then deciding what "scope of activity" can be permitted to aquatic organisms. Having decided what is required of a river or lake, then more purposive research can be carried out to determine the conditions needed to ensure its conservation for the required uses.

REFERENCES

Alabaster, J. S., and A. L. Downing. 1958. "The Behavior of Roach *(Rutilus rutilus* L.) in Temperature Gradients in a Large Outdoor Tank." In *Proceedings of Indo-Pacific Fisheries Council* 3:49.

Alabaster, J. S. 1963. "The Effect of Heated Effluents on Fish." *International Journal of Air and Water Pollution* 7:541.

Alexander, W. B., B. A. Southgate, and R. Bassindale. 1935. *Survey of the River Tees. Pt. II. The Estuary-Chemical and Biological.* Technical Paper, Water Pollution Research, H.M.S.O., London. 5.

Allanson, B. R., and R. G. Noble. 1964. "The Tolerance of Tilapia mossambica (Peters) to High Temperatures." *Transactions of the American Fisheries Society* 93(4):323–332.

Appourchaux, M. 1952. "Effets de la Temperature de l'eau sur la Faune et la Flore Aquatiques." *L'Eau* 8:377.

Bailey, R. M. 1955. "Differential Mortality from High Temperatures in a Mixed Population of Fishes in Southern Michigan." *Ecology* 36(3):526–528.

Baldwin, E. 1959. *Dynamic Aspects of Biochemistry.* Cambridge: At the University Press.

Baldwin, N. S. 1956. "Food Consumption and Growth of Brook Trout at Different Temperatures." *Transactions of American Fisheries Society* 86:323–328.

Banta, A. M., and T. R. Wood. 1928. "Genetic Evidence that the Cladocera Male is Diploid." *Science* 67:18.

Bishai, H. M. 1960. "Upper Lethal Temperatures for Larval Salmonids." *Journal Cons. Int. Explor. Mer.* 25(2):129–133.

Black, E. D. 1953. "Upper Lethal Temperatures of Some British Columbia Freshwater Fishes." *Journal Fisheries Research Board of Canada* 10:196.

Boycott, A. E. 1936. "The Habitats of Fresh-Water Mollusca in Britain." *Journal of Animal Ecology* 5(1):116–186.

Brett, J. R. 1956. "Some Principles in the Thermal Requirements of Fishes." *Quarterly Review of Biology* 31(2):75.

Brock, T. B. 1966. *Principles of Microbial Ecology.* New Jersey:Prentice-Hall.

Brown, V. M., D. H. M. Jordan, and B. A. Tiller. 1967. "The Effect of Temperature on the Acute Toxicity of Phenol to Rainbow Trout in Hard Water." *Water Research* 1:587.

Brues, C. T. 1939. "Studies on the Fauna of Some Thermal Springs in the Dutch

East Indies." In *Proceedings of the American Academy of Arts and Science* 73(4):71.

Bullock, T. H. 1955. "Compensation for Temperature in the Metabolism and Activity of Poikilotherms." *Biological Review* 30(3):311–342.

Cairns, J. 1956. "Effects of Heat on Fish." *Industrial Wastes* 18.

Cairns, J. 1955. "The Effects of Increased Temperatures upon Aquatic Organisms." In *Proceedings 10th Purdue Industrial Wastes Conference, Purdue University Engineering Bulletin* 40(1):346.

Cocking, A. W. 1959. "The Effects of High Temperatures on Roach (Rutilus rutilus)." *Journal of Experimental Biology* 36(1)203–226.

Collins, G. B. 1952. "Factors Influencing the Orientation of Migrating Anadromous Fishes." *Fish Bulletin of the U.S. Fish and Wildlife Service* 52:375.

Craigie, D. E. 1963. "An Effect of Water Hardness in the Thermal Resistance of the Rainbow Trout, Salmo Gairdneri Richardson." *Canadian Journal of Zoology* 41(5):825–830.

Creaser, C. W. 1930. "Relative Importance of Hydrogen-Ion Concentration, Temperature, Dissolved Oxygen and Carbon Dioxide Tension on Habitat Selection by Brook Trout." *Ecology* 11(2):246–262.

Davies, I. 1966. "Chemical Changes in Cooling Water Towers." *International Journal of Air and Water Pollution* 10:853.

Davies, L., and C. D. Smith. 1958. "The Distribution and Growth of Prosimulium larvae (Diptera Simuliidae) in Hill Streams in Northern England." *Jour. of Animal Ecology* 27:335.

Downing, K. M., and J. C. Merkens. 1957. "The Influence of Temperature on the Survival of Several Species of Fish in Low Tensions of Dissolved Oxygen." *Annals of Applied Biology* 45(2):261–267.

Ferguson, R. G. 1958. "The Preferred Temperature of Fish and their Midsummer Distribution in Temperate Lakes and Streams." *Journal of Fisheries Research Board of Canada* 15(4):607–624.

Fogg, G. E. 1965. *Algal Cultures and Phytoplankton Ecology.* Madison: University of Wisconsin Press.

Fowler, H. W. "List of Fishes Recorded from Pennsylvania." *Bulletin Commonwealth of Pennsylvania Board of Fish Commission.*

Fry, F. E. J. 1957. "Aquatic Respiration of Fish." In *The Physiology of Fishes: Metabolism,* Vol. 1, by M. E. Brown. New York: Academic Press.

Fry, F. E. J., J. R. Brett, and G. H. Clawson. 1942. "Lethal Limits of Temperature for Young Goldfish." *Revue Canadian Biology* 1(1):50.

Garside, E. T., and J. S. Tait. 1958. "Preferred Temperature of Rainbow Trout *(Salmo gairdneri* Richardson) and its Unusual Relationship to Acclimation Temperature." *Canadian Journal of Zoology* 36:563–567.

Gibson, E. S., and F. E. J. Fry. 1954. "Performance of Lake Trout, *Salvelinus namaycush,* at Various Temperatures and Oxygen Levels." *Canadian Journal of Zoology* 32(3):252–260.

Gotaas, H. B. 1948. "Effects of Temperature on Biochemical Oxidation of Sewage." *Sewage Works Journal* 20(3):441.

Harrison, M. E., and H. Heukelekian. 1958. "Slime Infestation—Literature Review." *Sewage and Industrial Wastes* 30(10):1278–1302.

Hart, J. S. 1947. "Lethal Temperature Relations of Certain Fish of the Toronto Region." *Transactions of the Royal Society of Canada,* 3d Series. Section V, 41: 57–71.

Hawkes, H. A. 1964. "An Ecological Basis for the Biological Assessment of River Pollution." *Chemistry and Industry* 437 (March).

Hawkes, H. A. 1963. "Biological Detection and Assessment of River Pollution." *Effluent and Water Treatment Journal* (Dec):651.

Hawkes, H. A. 1962. "Biological Aspects of River Pollution." In *River Pollution II. Causes and Effects,* by L. Klein. London: Butterworths.

Heilbrunn, L. V. 1955. *An Outline of General Physiology.* Philadelphia and London: Saunders.

Hoak, R. D. 1961. "Defining Thermal Pollution." *Power Engineering* 65(Dec):39–42.

Hoar, W. S. 1956. "Photoperiodism and Thermal Resistance of Goldfish." *Nature* 178(4529):364–365.

Hohnl G. 1956. "The Biology of Waste Water Fungi with Special Reference to Pulp and Paper Mill Effluents." *Wbl. Papierfabr.* 84:564.

Hubbs, C., and K. Strawn. 1957. "The Effects of Light and Temperature on the Fecundity of the Greenthroat Darter, *Etheostoma lepidum.*" *Ecology* 35(4):596.

Huntsman, A. G. 1942. "Death of Salmon and Trout with High Temperature." *Journal of Fishery Research Board of Canada* 5:485–501.

Hutchinson, G. E. 1967. *A Treatise on Limnology. Vol. 2.* New York, London, Sydney: Saunders.

Ide, F. P. 1935. "Effects of Temperature on the Distribution of the Mayfly Fauna of a Stream." *University of Toronto Studies, Biological Series* 39:9–76.

Laberge, R. H. 1959. "Thermal Discharges." *Water and Sewage Works* 106(Dec):536.

Lamanna, C., and M. F. Mallette. 1953. *Basic Bacteriology—Its Biological and Chemical Background.* Baltimore: Williams and Wilkins.

Lloyd, R., and D. W. M. Herbert. 1962. "The Effects of the Environment on the Toxicity of Poisons to Fish." *Journal of the Institute of Public Health Engineers* 132(July):132–145.

Long, Le Kuang, Nguem Din Zau, and Nguem Kuang Ving. 1963. "First Results of a Study of the Action of Temperature on the Fish, *Tilapia mossambica* Peters, Acclimated in Vietnam Since 1961." Translation, *Biological Abstracts* 41(9098):684.

Lund, J. W. G. 1950. "Studies on *Asterionella formosa* Hass 11. Nutrient Depletion and the Spring Maximum." *Journal of Ecology* 38:1.

Macan, T. T. 1963. *Freshwater Ecology.* London: Longmans.

Mann, K. H. 1965. "Heated Effluents and their Effects on the Invertebrate Fauna of Rivers." In *Proceedings of the Society for Water Treatment and Examination* 14:45.

Markowski, S. 1959. "Cooling Water of Power Stations. A New Factor in the Environment of Marine and Freshwater Invertebrates." *Journal of Animal Ecology* 28(2):243–255.

Markowski, S. 1960. "Observations on the Response of Some Benthic Organisms to Power Station Cooling." *Journal of Animal Ecology* 24(2):349–357.

Meuwis, A. L., and M. J. Heuts. 1957. "Temperature Dependence of Breathing Rate in Carp." *Biological Bulletin* 112(1):97–107.

Odum, E. P. 1959. *Fundamentals of Ecology.* Philadelphia and London: Saunders.

Patalas, K. 1967. "Original and Secondary Production of Plankton in the Lake Heated by Electricity Power Plant." *Summary Reports of 7th Hydrobiologists' Congress in Swinoujscie* (Sept. 1967).

Pearsall, W. H. 1923. "A Theory of Diatom Periodicity." *Journal of Ecology* 11:165.

Pitt, T. K., E. T. Garside, and R. L. Hepburn. 1956. "Temperature Selection of the Carp (*Cyprinus carpio* Linn.)." *Canadian Jour. of Zoology* 34:555–557.

Poltoracka-Sosnowska, J. 1967. "Composition by Species of Phytoplankton in the Lakes with Normal and Artificially Raised Temperatures." *Summary Reports on 7th Polish Hydrobiologists' Congress in Swinoujscie* (Sept. 1967)

Rawson, D. S. 1960. "A Limnological Comparison of Twelve Large Lakes in Northern Saskatchewan." *Limnology and Oceanography* 5(2):195–216.

Ross, F. F. 1959. "The Operation of Thermal Stations in Relation to Streams." *Journal and Proceedings of the Institute of Sewage Purification* 16.

Ruttner, F. 1937. "Limnologische Studien an Einigen Seen der Ostalpen." *Archiv Hydrobiology* 32:167.

Schlieper, C., and J. Blasing. 1952. "Uber Unterschiede in dem individuellen und okologischen Temperaturbereich von *Planaria alpina* Dana." *Arch. Hydrobiology* 47:288.

Smyk, B., and M. Cienciala. 1959. "Wplys Sciekow przemyslu chemicznego i energetycznego no stan mikrobiologiczny wod rzek gorskich." *Acta Microbiology, Polonica* 8:125.

Spaas, J. T. 1961. "Contribution to the Comparative Physiology and Genetics of the European Salmonidae. III Temperature Resistance at Different Ages." *Hydrobiologia* 15(1–2):78–88.

Sprague, J. B. 1963. "Resistance of Four Freshwater Crustaceans to Lethal High Temperature and Low Oxygen." *Journal of Fisheries Research Board, Canada* 20(2):388–415.

Stangenberg, M., and M. Pawlaczyk. 1960. "Wplyw Zrzutu Wod. Cieplych Z Elektrowni na Ksztaltowanie sie Biocenozy Rzecznej." *Zeszyty Naukowe Politechniki Wroclawskiej, Inzynieria Sanitarna* 40:67.

Sullivan, C. M., and K. C. Fisher. 1953. "Seasonal Fluctuations in the Selected Temperature of Speckled Trout *Salvelinus fontinalis* (Mitchell)." *Journal of Fisheries Research Board of Canada* 10:187.

Sullivan, C. M. 1954. "Temperature Reception and Responses in Fish." *Journal of Fisheries Research Board, Canada* 11(2):153–170.

Tarzwell, C. M. 1956. "Water Quality Criteria for Aquatic Life." In *Transactions of a Seminar on Biological Problems in Water Pollution*. Cincinnati: Robert A. Taft Sanitary Engineering Center.

Trembley, F. J. 1960. *Research Project on Effects of Condenser Discharge Water on Aquatic Life. Progress Report, 1956–1959*. Bethlehem, Pa.: Lehigh University Institute of Research.

Truesdale, G. A., and K. G. Vandyke. 1958. "The Effect of Temperature on the Aeration of Flowing Water." *Water and Waste Treatment Journal* 9 (May-June).

Varley, M. E. 1967. *British Freshwater Fishes*. London: Fishing News (Books) Ltd.

Walshe, B. M. 1948. "The Oxygen Requirements and Thermal Resistance of Chironmid Larvae from Flowing and from Still Waters." *Journal of Experimental Biology* 25:35.

Weatherley, A. H. 1959. "Some Features of the Biology of the Tench *Tinca tinca* (Linnaeus) in Tasmania." *Journal of Animal Ecology* 28(1):73–87.

Weatherley, A. H. 1963. "Thermal Stress and Interrenal Tissue in the Perch *Perca fluviatilis* (Linnaeus)." *Proceedings, Zoological Society, London* 141(3):527–555.

Webster, D. A. 1945. "Relation of Temperature to Survival and Incubation of the Eggs of Smallmouth Bass (Micropterus dolomieu)." *Transactions of American Fisheries Society* 75:43–47.

Wesenberg-Lund, C. *Plankton Investigations of the Danish Lakes*. Special Par. Copenhagen.

Wuhrmann, K., and H. Woker. 1953. "Contributions to the Toxicology of Fish. VIII.

The Toxicity to Fish of Solutions of Ammonia and Cyanide at Different Oxygen Tensions and Temperatures." *Schweiz, Z. Hydrol.* 15:235.

Wurtz, C. B., and C. E. Renn. 1964. *Water Temperatures and Aquatic Life—Prepared for Edison Electric Institute, Research Project No. 49.* Baltimore: Johns Hopkins University.

Wurtz, C. B. 1961. "Is Heat a New Pollution Threat?" *Wastes Engineering* 32(2):684.

DISCUSSION/ Eugene B. Welch

MR. HAWKES has done an excellent job of describing the potential and significant effects from heated water, so rather than reiterate on the various points, I would rather stress the main theme, that of interpreting biological effects in terms of their applied significance. I would like to do this by describing some specific biological data related to raised water temperature in the Tennessee Valley and some planned research by TVA.

Approaches to the study of thermal problems and interpretation of results are somewhat variable, depending on the position of responsibility of the investigator. The approach and interpretation may be quite different if the objective is for maximum use of a stream or less than maximum use or even for a single use. To attain maximum use requires determination of the maximum heat load tolerated without significantly reducing the desired biological populations.

Any appreciable change in the temperature regime of an environment will no doubt produce some effect on the biological community. The first problem, however, is to detect the effect biologically and the second is to determine whether the effect is beneficial or detrimental to the biological community and to what extent. The ultimate question is, how do the measured effects quantitatively relate to the desired or economically important species?

If maximum use of the water is desired, then a decision is necessary relative to how much detrimental biological effect will be tolerated. To obtain data to allow such a decision, different and more elaborate studies than have been made, to date, will be necessary. Ideally, it would be desirable to know the relationship between annual production of economically important species and increments of added heat. Of course, biological field measurements are not sufficiently refined to provide such answers readily, and the results of laboratory studies, where the precision is attainable,

58

cannot be reliably applied to the environment because most of the inter-acting factors have been removed or held constant in the laboratory.

To obtain some of the needed answers to such questions as how much heat is allowable while an acceptable level of production of economically important species is still maintained and relationships between such pro-duction and appropriate biological indices are developed, experimental re-sults are needed from semicontrolled environments where as many natural environmental factors as possible are allowed to operate, but where the heat addition can be rigidly controlled. Such results could be readily applied to the natural environment with a minimum of judgment necessary. In the immediate future, however, until such ideal data are available, interpreta-tion of existing field and laboratory data will be necessary in order to esti-mate the extent and significance of biological effects. Observed and potential biological effects of heated water in the Tennessee Valley are described to indicate problems of data interpretation, and planned research in TVA to provide the needed experimental results are briefly outlined.

PROBLEM OF DATA INTERPRETATION

Since biological populations are rather sensitive to environmental change, addition of heated water is likely to produce a biological change. Although this change is probably detectable, it becomes more important to interpret such a change in terms of the total ecosystem or economically important species.

Biological effects from raised temperature in the Green River, near the Paradise Steam Plant in Kentucky, represent such a problem of interpreta-tion. The periphyton (attached organisms, or "slimes") community shows a response to temperature in this area, as indicated by growth accumulated on plexiglass slides submerged 1.5 feet below stream surface for four 2-week periods in 1965 and six 2-week periods in 1966. Figure 1 shows that the means of daily maximum and minimum temperatures during these periods of study were much greater in 1966, a low-flow year, than in 1965, a rela-tively high-flow year. Correspondingly, the mean autotrophic indices, or simply the ratio of phytopigment content to total weight of the periphyton, was much more depressed in the area of heated water in 1966 than in 1965. This might be interpreted as reduced production of autotrophic relative to heterotrophic organisms. However, total accumulation of organic matter (mean ash-free-dry-weight) was much greater at nearly all stations in 1966, with the peak occurring at river-mile 94.5, five miles below the heated-

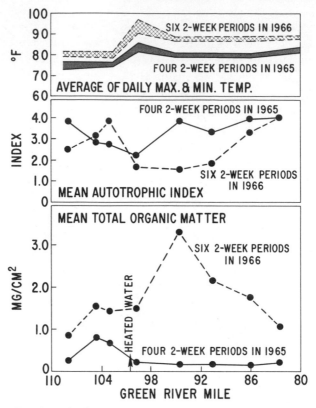

Fig. 1—. Surface temperature, autotrophic index, and accumulation of organic matter near the Paradise Steam Plant, Green River, Ky.

water discharge—clearly an effect of increased temperature during that year. This peak in growth at mile 94.5 is thought to be a response to more rapid cycling and release of nutrient material in this area, rather than a response to a more optimum temperature. Growth should have remained high at all stations further downstream if solely a response to temperature. Instead, growth steadily declined downstream from river-mile 94.5, where the peak occurred.

The variation of selected water-quality characteristics supports the contention that the maximum growth of organic matter five miles below the heated-water discharge is a result of increased cycling and release of nutrients. Figure 2 shows that, coincident with the temperature increase, ammonia content increased greatly and soluble phosphate increased slightly. These values are means of determinations in samples collected at surface, middle, and bottom depths on July 26, 1966. A slight downstream depres-

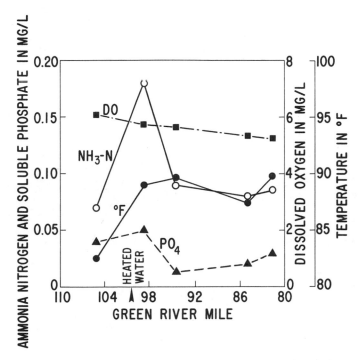

Fig. 2——. Water temperature, ammonia, soluble phosphate, and dissolved oxygen near the Paradise Steam Plant, Green River, Ky., on July 26, 1966.

sion in oxygen content occurred, which also may be indicative of increased rates of nutrient cycling. The decrease is not due to physical loss from heated water that is supersaturated with DO, because saturation did not exceed 100 percent.

Although the increase in production of periphyton organic matter is probably an indirect result of raised temperature, the magnitude was not sufficient to cause a nuisance problem. But can the increased production be considered beneficial? That is, has production of fish food and fish also increased?

The Green River is rather deep, slow-moving, and the surface elevation is regulated for navigation. Therefore, littoral area and benthic invertebrate habitat are scarce, so that zooplankton is probably the base of the food chain in that environment. As a result, zooplankton are considered an important indication of temperature effects, particularly as they relate to the economically important fish populations. Experiments showed that the lethal temperature of the entomostracan plankters was about 96° F., regardless

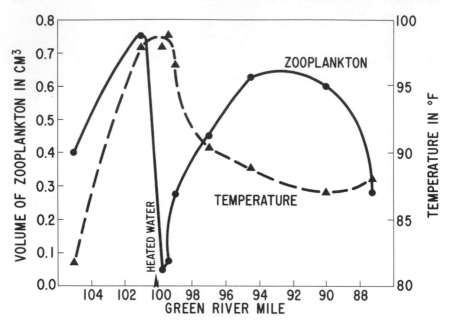

Fig. 3——. Zooplankton volume and water temperature near the Paradise Steam
Plant, Green River, Ky., May 26–27, 1964.

of the rate of acclimation. Therefore, when the temperature of condenser
cooling water exceeded that lethal temperature and when an appreciable
portion of the riverflow was used for cooling, a significant reduction in the
zooplankton stock occurred. Figure 3 shows the catch from 5-minute surface tows with a net in relation to temperature at the indicated stations in
May 1964. The reduction in zooplankton volume immediately below the
heated discharge resulted from raising the temperature of the water passing
through the condensers to some temperature above the lethal limit. The
importance of allowing an appreciable portion of the flow to pass by the
intake and contribute plankton stock downstream was recognized early in
plant operation. As Figure 3 shows, zooplankton were abundant upstream
from the discharge at temperatures near the lethal limit. Following the
decrease in stock immediately below the discharge, populations of adult
zooplankton returned within three to five miles downstream to levels that
were comparable to and even exceeded levels at the upstream control. Adult
copepods and cladocerans alike were largely eliminated in the area of the
heated water discharge, but rotifers and copepod larvae were only slightly
reduced. Copepod larvae returned to control levels within a mile of the
discharge and more than doubled control levels within 5 miles of the dis-

charge. The decrease in zooplankton stock at mile 87.2 is due primarily to an unexplained lack of cladocerans, but copepods remained comparatively high.

Although there was a decided decrease in zooplankton stock immediately below the steam plant, production was no doubt increased downstream, as indicated by a greater proportion of larval forms in samples, possibly in response to an increased nutrient recycling and production of particulate food. This was suggested earlier as an explanation for the greater accumulation rate of periphyton organic matter about five miles downstream from the heated discharge. The decrease in nutrient content downstream from mile 94.5 (Figure 2) may indicate that it was tied up by the increased zooplankton stock. The detrital material originating from dead plankton in the coolant water may contribute to increased production of heterotrophic organisms and may be used as food by zooplankton. Thus, in actuality, has the total net production of fish food decreased?

Results of winter fish-netting in the area of heated water discharge from 1961 to 1967 shows a progressive and large increase in total catch, as well as an increase in the proportion of game fish and reduction of rough fish in the catch. Even if fish avoid the heated-water area during the warm months, the question nevertheless remains: Should the biological effects in this instance be interpreted as beneficial or detrimental? Clearly, data are needed on the long-term biological effects of raised temperature which are related to economically important species and their supporting food organisms and can be readily applied to determine temperature standards.

DETERMINATION OF CRITERIA

As desirable as it might be to maintain natural biological structure and production and still use the water for extensive cooling, it is probably not possible. A highly developed river, such as the Tennessee, has already been changed considerably from the natural condition, so striving for such a "natural" condition seems unrealistic. What does seem realistic is to maintain temperature standards that insure "adequate" reproduction and growth of economically important species. The biologist, then, must provide criteria that are necessary to produce an adequate crop of desirable organisms. Such criteria must be based on actual field conditions—that is, the daily and seasonal temperature regimes that the aquatic organisms are exposed to.

Potential Effects of Added Heat

As a guide to the development of criteria, it is instructive to consider the

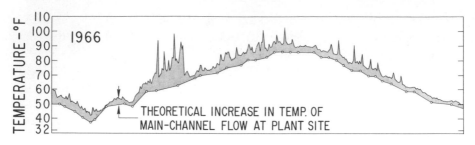

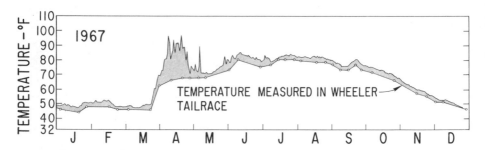

Fig. 4——. Theoretical temperature in Wheeler Reservoir below the Browns Ferry
Nuclear Plant.

potential annual temperature regimes actually expected below a power
plant cooling-water discharge. For example, the condenser cooling-water
system for the Browns Ferry Nuclear-Power Plant is designed to limit the
mixed water temperature below the plant to a rise of 10° F. and a maximum
of 93° F. Figure 4 shows temperatures observed in the tailrace below
Wheeler Dam during 1966 and 1967, along with hypothetical daily temperatures that would occur 19 miles upstream, immediately below the
Browns Ferry Nuclear Power-Plant, if coolant water from all three units
(4400 cfs) of the power plant were continually added to Wheeler Reservoir
and mixed with two-thirds of the daily flow (two-thirds of the total flow is
in the old river channel) as measured downstream at the dam. The long-
term mean flow at the dam is 49,000 cfs. The mean flow in 1966 was less
than this value (37,000 cfs), but was greater (57,000 cfs) in 1967. The
results show that the mixed temperature below the steam plant of two-
thirds of the total flow would have reached or exceeded a rise of 10° F.
above ambient only 44 days in 1966 and 25 days in 1967. A maximum of
93° F. would have been reached or exceeded only 15 days in 1966 and 3
days in 1967 (TVA, 1968). Thus, special flow regulations will be necessary
for only a small percentage of the time in order to meet the design-tem-

perature standards. However, the significant aspect in Figure 4 is the large percentage of time that the rise in temperature would be much less than the allowable 10° F., i.e., not more than 3° or 4° F. Of course, this is because, most of the time, normal flows will minimize the temperature rise. This factor must be considered when determining criteria by extrapolation from experimental results to field conditions. That is, not only should it be necessary to know the upper tolerance of thermal rise for the most sensitive life-history stages of important species, but also when and for how long are such limits required.

Maintenance of the maximum summer temperature of the river below 93° F., after mixing of the heated discharge, is considered a reasonable standard for protection of warm-water fisheries. This is below determined lethal limits for most economically important warm-water species; however, it may be too high for some of the minnows. The sunfishes and basses have lethal limits considerably above 93° F., while catfishes are a little less tolerant than sunfishes (Hart, 1947; Brett, 1956; Cairns, 1956; Trembley, 1960). However, survival depends on the period of exposure to a given temperature the length of photoperiod, and degree of acclimation and not only on a measured temperature *per se*. The shorter the exposure time, the higher are the maximums that can be tolerated up to an ultimate limit (Alabaster and Downing, 1966). An organism is adapted to a normal daily and seasonal pattern of fluctuating temperature. During relatively high mid-day temperatures, the organisms can compensate metabolically if lower night-time temperatures are encountered. Since many of the warm-water species acclimate rather fast to fluctuating temperature, particularly to rising temperature (Cairns, 1956; Doudoroff, 1957; Nicum, 1967), the greater the diurnal range, the greater is the maximum that will be tolerated. Thus, lethal temperatures should be determined for all important species in a particular environment incorporating the normal diurnal range in temperature.

The length of time and season that near-maximum temperatures are encountered can be expected to affect reproduction and growth of the organisms. A limit on increment rise above ambient insures that annual maximum temperatures will be reached only during the summer, and not in the breeding season. Thus, critical egg and larval stages are not exposed to maximum temperatures that would surely interfere with normal development. The potentially high temperatures that would be encountered during April (Figure 4), the period of upstream-reservoir filling, will be controlled by special flow-regulation to stay within the 10° F. rise. However, the period of exposure to normal annual maximum temperatures may be important. Note that without the addition of heated water, the annual maximum of 85° F. occurred for less than a month during each year; but with heated water, the period in which temperatures would equal or exceed 85° F. would increase to two and one-half or three months. Since organisms grow best at their optimum or preferred temperatures for activity (Brett, 1956), for which, for most warm-water species, 85° F is at or above the upper end of the range, such a changed temperature regime may decrease growth. The significance of this factor is yet to be determined.

A maximum allowable temperature rise above ambient is much more difficult to determine with existing knowledge than a lethal limit because experimental results on many important species are nonexistent and, for the most part, incomplete on those that have been studied. However, there is sufficient evidence to indicate that lower winter temperatures are necessary for normal development and proliferation of the gonads in some warm-water species (Hubbs and Strawn, 1957; Merriman and Schedl, 1941) and that a gradual increase in mean daily temperatures (along with photoperiod) in the spring is necessary for initiation of spawning and successful development of eggs and larvae (Volodin, 1960; Lillelund, 1966). Certainly this life-history stage is most critical for the majority of species. Thus, it becomes important to consider flow regulations to minimize rapid day-to-day temperature fluctuations characterized by the low-flow condition, as shown during April in Figure 4.

Planned Research by TVA

TVA is developing plans for large-scale experiments to determine the tolerance of selected groups of biological populations to heated-water discharges. The results of such research should be of widespread interest. Funding of the project on a basis compatible to this breadth of interest is to be sought.

The projected studies would require construction of eight channels, 16 feet wide, approximately 400 feet long, and ranging in depth from 1 to 4 feet. A maximum of 9 cfs of water would be supplied to each channel, and the option of microstraining the water to remove zooplankton ($>100\mu$) would be available. The physical and chemical environment among channels would be maintained as similar as possible, except that temperature would be automatically controlled. Heated water from the condenser discharge would be added to natural river water of ambient temperature, to provide desired experimental regimes of constant or fluctuating temperature. The site under study for the experimental channels is TVA's Browns Ferry Nuclear-Power Plant now under construction. Subject to timely funding of the project, the channels would be constructed and ready for initiation of controlled experiments soon after the power station goes into operation in the fall of 1970.

Initially, the relationship between biological production and increment rise in temperature above ambient would be studied. Two channels would receive unheated river water as controls and each of the three remaining pairs of channels would receive increment rises above ambient of 5°, 10°, and 15°, respectively. Maximum water-temperature rise in the power plant condensers is 25° F. Channel temperatures would not exceed a maximum of 93° F. The productivity at each level of the food chain would be determined in each channel, which should result in replicated estimates of annual production at three test-temperature regimes and a control of normal river temperature. This experimental design should test only the effect of temperature, since particulate matter would be screened out; and populations in each channel should be supported on the food supply produced within that channel. After the relationship between a constant rise above ambient and biological production is established for one or more groups of populations, which may take from two to five years, the effect of various patterns of fluctuation of a given rise above ambient would be tested.

Preliminary tests would be conducted in the laboratory to "screen" species of commercial, game, and forage fish for the long-term channel experiments. This procedure is necessary to determine the most compatible species combinations based on their maximum-temperature tolerance for survival and growth under laboratory conditions. Nine species studied in the laboratory, from which representatives for the channel experiments would be selected, are white crappie, largemouth bass, channel catfish, bluegill, smallmouth bass, gizzard shad, lake-emerald shiner, roseyfaced shiner, and white bass. The temperature tolerance of these fishes would be studied in four seasonal sequences with each life-history stage represented during each of the four seasons. The channels would receive river water of normal temperature for one year prior to the availability of heated water in 1970. During this time, populations of selected fish and naturally recruited fish-food organisms should become established.

The effect of temperature on fish growth and production in the vicinity of heated-water discharges is not well known. This is largely because sample collection of particular fish with known temperature histories is difficult in the natural environment, since fish tend to avoid temperatures out of their preferred range. In the experimental channels, the organisms would have no choice but to remain exposed to the predetermined temperature regimes. Such an experiment should provide the natural interaction between such characteristics as an increased growing season, changed reproductive timing, and greater metabolic expenditures for activity. Allowing all these conditions to act simultaneously would provide a resulting production that can be related to the increment of temperature rise above ambient and should, therefore, reliably forecast conditions that would exist in the natural environment.

REFERENCES

Alabaster, J. S., and A. L. Downing. 1966. "A Field and Laboratory Investigation of the Effects of Heated Effluents on Fish." *Ministry of Agriculture, Fisheries, and Food, Fishery Investigations* 1 (6):1–42.

Brett, J. R. 1956. "Some Principles in the Thermal Requirements of Fishes." *Quarterly Review of Biology* 31:75–87.

Cairns, J., Jr. 1956. "Effects of Heat on Fish." *Industrial Wastes* 1:180–183.

Doudoroff, P. 1957. "Water Quality Requirements of Fishes and Effects of Toxic Substances." In *The Physiology of Fishes,* edited by M. E. Brown, vol. 2, pp. 407–408. New York: Academic Press, Inc.

Hart, J. S. 1947. "Lethal Temperature Relations of Certain Fish of the Toronto Region." *Transactions, Royal Society, Canada* 41:57–71.

Hubbs, C., and K. Strawn. 1957. "The Effects of Light and Temperature on the Fecundity of the Greenthroat Darter, *Etheostoma lepidum.*" *Ecology* 38:596–602.

Lillelund, K. 1966. 'Versuche zur erbrütung der eier vom Hecht, *Esox lucius L.,* in Abhangigkeit von Temperatur and Licht." *Archives fur Fischereiwissenschaft* 17:95–113.

Merriman, D., and H. P. Schedl. 1941. "The Effects of Light and Temperature on Gametogenesis in the Four-Spined Stickleback, Apeletes quadracus (Mitchell)." *Journal Experimental Zoology* 88:413–449.

Nicum, J. G. 1967. "Some Effects of Sudden Temperature Changes Upon Selected Species of Freshwater Fishes." Thesis, Southern Illinois University.

Tennessee Valley Authority. 1968. "Plans for Water Temperature and Radwaste Control, Browns Ferry Nuclear-Power Plant—Wheeler Reservoir." (June):9.

Trembley, F. J. 1960. "Research Project on Effects of Condenser Discharge Water on Aquatic Life." *Progress Report, 1956–1959.* Bethlehem, Pa.: Lehigh University Institute of Research.

Volodin, V. M. 1960. "Effect of Temperature on the Embryonic Development of the Pike, the Blue Bream (Abramis Ballevus L.) and the White Bream (Blicia Bjoerkna L.)." *Trudy Institute Biologii Bodskhranilisch* 3:231–237.

DISCUSSION FROM THE FLOOR

John Foerster: I am a microbiologist. We have seen in the two papers the possibility of thermal pollution being an asset. Information has been presented to us on the basis of primary productivity, autotrophic indexes, diversity in the species as far as the Polish lakes were concerned, where the warm-water lake had more species than the cold-water lake.

However, one thing I think engineering people should be made aware of is that diversity indexes and primary productivity don't mean a tinker's dam when you are talking about food chains, because large primary productivity can be contributed by blue-green algae, which have little or no importance in the food chains.

If the authors would like to make a comment on this, I would appreciate it.

Also, the Polish lakes studied by Dr. Hawkes did not seem to receive appreciable amounts of heat, 88° F. being the maximum, I think, of the warm-water lake, which seemed to be pretty deficient, as far as biology goes.

In the work that we are involved with on the Connecticut River in regard to the Connecticut Yankee project, we now have river temperatures exceeding 103° F.; the fish are dead; and it stinks like a sewer. You wonder whether these primary productivity indices mean much because you do get organisms that grow in this stuff and they do seem to produce well. But they aren't being utilized. I would appreciate your comments on this.

H. A. Hawkes: I agree fully with the speaker about primary productivity. It depends on whether you want what you are producing or not. In many cases, the changes being observed are in blue-green algae which, up to now, I have understood is less acceptable and probably a nuisance value, rather than of any benefit.

69

However, some recent work I have been discussing with people in the British Biological Association seems to indicate that primary productivity, in terms of the blue-green algae, may result in a bacterial increase which is used by the consumer population. Whether these organisms can use the bacteria produced from the decay of the blue-greens, I am not sure.

In the case of the Polish lakes, whether things are going to be beneficial or detrimental depends upon the temperature range in which we are working, rather than the actual temperature increase.

I didn't emphasize sufficiently that temperature can be limited in either system to the low level as the high. I know in a symposium on thermal pollution we are all thinking of the high temperatures being a limit. Low temperatures are also a limiting factor. Where this is so, then the elevation of this temperature will bring about increased productivity as it did in the Polish lakes.

Eugene B. Welch: In terms of the Green River, we looked at the species diversity there, and certainly there were some changes in species diversity.

Voice: Dr. Welch, you said during April and early May you would have a considerable rise in temperature. Now, how would this go, according to the river standards which are proposed for the area or which are likely to be applied to the area?

In short, would the plant have to shut down during these periods?

Welch: No. As I mentioned, there will be special upstream water releases to maintain the temperature rises within the standards, so this condition is strictly hypothetical.

Tom Wright: I would like to see if I could get Dr. Welch to elaborate a bit on this concept of multiple use. It is a quasi-biological type of term which seems to promise a chicken every Sunday and something for everyone, usually at no cost.

This has been used, as many of you are aware, to justify all manner of things. It has been used to justify dams where they are no longer needed for hydroelectric projects, on the basis that they are useful for recreation, irrigation, esthetics, any number of other things. I would cite some of the more recent ones in the Far West as examples of this.

In this concept, I question who will decide, under multiple use, who is going to have this use, whether it will be industry, the general public, the particular segment of the public, or some other special-interest group, and how these various uses will be balanced out.

For example, should we shift the balance of a lake which is now a bass lake to a carp lake, which, believe it or not, will please many fishermen—lower-class fishermen—without raising the temperature which will have implication for perhaps a hydroelectric facility and so on?

Would you care to comment on this, the ramifications of multiple use?

Welch: I probably can't elaborate any more than you did, except that in this position of trying to interpret these results, one comes under very questionable situations, I think.

William Davis: If I understand the TVA concept, the biota of natural waters should be modified to adapt to industrial effluents rather than the adaptation of industrial effluents to natural water and production of natural biota.

I want to point out that we have had terrific problems controlling the environment in order to maintain productivity. We have been successful with trout and salmon, but this requires a complete control of the total environment and it is extremely difficult to talk about changing or depositing waste materials in a natural environment where there are a terrific number of factors over which you have no control, and in managing this system, so that the natural biota will not multiply appreciably.

Welch: This is certainly true. I don't think we are geared right now to manage a situation that is as intensive as this. In terms of generalizing on the thing, it comes down to very minor changes in standards.

My only argument is for a better method by which to interpret standards. I think a biologist puts himself in a difficult position coming out for standards without more data than we have now to back him up.

WATER-QUALITY STANDARDS FOR TEMPERATURE

THIS symposium is one more example of the growing concern about thermal pollution, particularly as it relates to the damage of aquatic life. In fact, while water temperature has some impact on all major water uses, thereby allowing a definition of thermal pollution as any change in natural water temperature that adversely affects the aquatic environment, I think we can narrow the subject matter for discussion today by focusing on the requirements for fish and other aquatic life. Certainly this water use is much more sensitive to temperature increases than are water uses such as municipal and industrial water supplies. The water-quality standards program is designed to protect beneficial uses of interstate (including coastal) waters through application of numerical and narrative limits on pollutants, and the specification of necessary treatment and control measures. Standards represent the basic strategy for controlling all water pollution. The legal basis for the standards approach is the Water Quality Act of 1965. Since its passage, all fifty states have adopted standards and, as of today, thirty-seven have received approval by the Secretary of the Interior in accordance with the Act.

When considering the phenomenal growth of electric power production in recent years, it is reasonable to say that the Act and the resultant standards were developed none too soon. It is my understanding that the power demand doubles every six to ten years and that, by 1980, the power industry will require perhaps 25 percent of the total available freshwater runoff in the entire nation for cooling purposes. The waste heat resulting from this increasing power demand must be controlled, along with other waste discharges, if a pleasant and productive water environment is to be preserved. In some locations, the capacity of the aquatic environment to absorb heat without suffering damage is being exceeded; and in many others, the situation is becoming critical. Water-quality standards represent our best hope to manage the aquatic environment; all contain temperature criteria to protect aquatic life.

Temperature criteria were developed with the basic objective of protecting native aquatic life by limiting artificial changes in the aquatic environment so that optimum rather than marginal conditions prevail. Unchecked waste-heat discharges which raise water temperatures several degrees may seriously alter the ecology of a lake, a stream, or even part of the sea. The biota in these waters is the result of long evolutionary processes, during which delicate balances were established; a small man-made change in conditions can have far-reaching effects.

Aquatic organisms are affected directly by temperature increases but, in addition, they are affected indirectly through temperature effects on other forms of aquatic life which comprise their food, competitors, and predators. Effective standards must recognize this complication. Furthermore, because of the increased sensitivity of aquatic life, artificial changes in water temperature should be particularly limited if other quality parameters, such as dissolved oxygen and toxic materials, are unfavorable.

Development of Temperature Standards

All fifty states submitted water-quality standards containing temperature criteria. Some noticeable *characteristics* of the criteria are:

A. They vary significantly from state to state.

B. Both numerical and narrative approaches were used by most states.

C. Numerical criteria were tailored to the type of fishery to be protected.

D. Numerical limits generally referenced a seasonal maximum, allowable change from background temperatures, and occasionally a rate of change.

E. Some reference was made to *mixing zones*.

The diverse temperature criteria exemplify, on the one hand, the commitment required to control thermal pollution and, on the other hand, the difficulty experienced by the states in knowing what was an appropriate numerical limit for temperature. A number of problems were faced in setting limits for this important but complicated water-quality parameter.

There is, for example, a lack of data on existing or natural temperatures. Where data are available, they are often scattered among several collecting agencies or are hidden in records, and they have not been fully evaluated or verified as to accuracy or applicability. For many waters, little or no valid data are available. Knowledge of existing or natural temperatures is necessary, both as a foundation for establishing criteria which are appropriate to local conditions, and as a basis for implementing and enforcing the criteria.

We have limited information on effects of heat loads. There is inadequate identification of aquatic life inhabiting particular waters, as well as inadequate knowledge of temperature tolerance of aquatic life. A good deal of this information is probably available, but it is scattered among the various state and federal agencies having water-pollution control and fisheries responsibilities and will take time to accumulate and evaluate. The area where information is particularly lacking is the effects of temperature on a number of important species at various life stages. Behavior responses involved in reproduction, migration, competition, and predator-prey relationships are virtually unknown, as are the effects of temperature change on susceptibility to parasites and pathogens.

There is uncertainty as to the best way to administer temperature criteria. Tying criteria to normal or natural conditions begs the question of what is "natural," where does "natural" apply, and what are appropriate mixing zones. There are problems in treating the occurrence of cumulative temperature increases, of separating and dealing with adverse effects of natural and man-made influences, and of providing criteria that are sufficiently flexible to accommodate seasonal variables.

Notwithstanding the many difficulties resulting from the lack of firm information and guidance, temperature criteria have been adopted that generally provide protection from heated-water discharges potentially injurious to aquatic life processes.

A frequent reference used by the states in adopting standards was the criteria developed by the Ohio River Valley Water Sanitation Committee. For the protection of a warm-water fishery, these criteria call for a summer maximum of 93° F. and a winter maximum of 73° F.

Furthermore, it is recommended that no temperature increase be allowed in streams suitable for trout propagation. There is no numerical limit on temperature change specified. Recommendations of other interstate groups and state fishery agencies are also reflected in the criteria adopted by individual states.

After June 30, 1967, the interim recommendations of the National Technical Advisory Committee (NTAC) were available, and these were used in our negotiations with the states, where we sought to upgrade the standards before approval was recommended to Secretary Udall, or where certain refinements appeared desirable in approved or generally approvable standards. The NTAC recommendations call primarily for adherence to "natural" temperatures within a narrow range. This is translated into a recommended limit of 5° F. artificial increase for cold and warm freshwater fisheries, except that no heated effluents should be discharged into spawning areas, headwaters of trout and salmon streams, and the hypolimnion of lakes and reservoirs. For marine waters, the recommended artificial-temperature change limits are 1.5° F. for the three summer months and 4° F. for the remainder of the year.

The objective of limiting the deviation from background temperatures is to preserve the normal daily and seasonal temperature variations that were present before the addition of heat due to other than natural causes. These criteria are complemented by the NTAC-recommended maximum temperatures for various species of fish. These values range from 48° F. (spawning and egg development of lake trout and salmon) to 93° F. (growth of catfish, certain types of bass, and other fish). The Committee emphasizes that these limits are considered to be "satisfactory" for aquatic life; they are not necessarily the optimum conditions.

The individual state standards include a narrative statement limiting temperature increase to levels not having deleterious effects on beneficial water uses. In addition, all but one state adopted numerical temperature limits. These numerical criteria all contained a maximum temperature limit which, in the forty-nine states, varied from about 55° F. to 96° F. Most contained a limit on allowable changes from background or natural temperatures. These varied from no increase at all (for trout waters) to as much as 20° F. Most were in the range from 4° F. to 10° F. change from "natural."

In addition, a number of states adopted a rate-of-change temperature criterion to protect aquatic life from a sudden sharp change in their environment. The value most frequently used is a limit of 2° F. change per hour.

Department of the Interior Reaction to Temperature Standards

The Department has sought to approve as many standards as possible without undue delay. We believed that it was better to get the nationwide effort to install treatment and control facilities under way as soon as possible, accompanied by the best criteria that could be developed with the information and time available.

In our negotiations with the states, we got a firm commitment to prevent and abate thermal pollution, but a mixed lot of criteria. We attempted to assure that approved criteria were compatible at state borders; that they embodied the principles of the NTAC and were reasonably compatible with its specific recommendations, as well as reasonably compatible with what information was available about existing water quality and aquatic life in specific areas. We were able to negotiate modifications in most cases that met these objectives. However, some states were not willing to adopt criteria consistent with the NTAC recommendations and therefore, while it is recognized that the recommendations are general and not tailored to conditions in specific streams, exceptions to the general approval of water-quality standards have been made for temperature criteria. In these instances, we are sitting down with the states to evaluate available data with the objective of developing criteria more in line with existing water quality and the NTAC recommendations. The basic reasons for the rejection of state-adopted temperature criteria have been: (1) too lenient limits on allowable man-made changes from natural background temperatures, and (2) maximum temperature limits incompatible with existing temperature maximums.

Administration of Temperature Standards

Even though the states, under the law, had the first opportunity to set the standards, this has been from the beginning a joint state-federal effort. As standards are implemented, the state-federal partnership will continue. We anticipate working together to monitor and evaluate compliance with standards and to make appropriate revisions in temperature criteria, as well as other parameters.

Monitoring compliance will require a recognition that standards are designed to protect aquatic life at extreme, not average, conditions. Also, implementing standards should be considered in relation to the cumulative effect of all projected heat inputs. The electric-power industry has years of experience in accurately projecting power needs for many years in the future. Our position, as regulatory agencies, must recognize the fact that there will be a four-fold expansion in power output over the next 20 years. In many parts of the country, this expansion will be accompanied by changing patterns of river flows, which will complicate the administration of temperature standards.

All of the water-quality standards discuss mixing zones, the context being that criteria do not apply within such zones except in those few cases where separate criteria were established for mixing zones and areas outside of mixing zones. Mixing zones should be established by proper administrative authority; it is expected that the size of the zones will vary with the location, size, character, and use of the receiving water. By definition, a mixing zone could be an area where unfavorable conditions exist for aquatic life. Therefore, it should be kept as small as possible through the use of good engineering design techniques and not be used as a substitute for good waste treatment and control.

Areas of mixing are unavoidable, but in all cases they should provide for passage of fish and the food that fish depend on for survival. The NTAC recommends a passageway containing 75 percent of the cross-sectional area and/or volume of flow of the stream or estuary. If there are several mixing areas close together, they should all be on the same side so that the passageway is continuous.

In summary, water-quality standards are a way, we feel, of coming to grips with thermal-pollution problems by protecting the ecological balance which nature has been able to maintain in spite of natural causes which frequently result in large variations in water temperature on a seasonal and even on a daily basis. Standards are the basic vehicle through which judgments are made concerning heat discharges. But the success of the standards approach will depend on cooperation between all parties concerned. The cooperative approach should include a number of factors that are important in administering standards and in future refinements of standards. These include:

A. More attention to site location.
B. Increased attention to the long-range use of power as it effects peak versus base loads, flow regulation, and the total management of river systems.
C. Acquiring a better understanding of the effects of heat on the aquatic environment.

The technology exists to comply with good temperature standards and the economics are feasible. Therefore, the term *thermal pollution* should become associated strictly with prevention, rather than abatement of an existing situation.

DISCUSSION/ Milo A. Churchill

SINCE a copy of Mr. Burd's paper was not received in time for consideration in preparing these comments, the writer is limited to a discussion of the subject itself—modified, to some extent, to take into account the apparent philosophy of the FWPCA concerning water-quality standards for temperature.

Since quality standards for those waters used by aquatic life are the most restrictive as regards addition of heat from man-made sources, this discussion will be limited to temperature standards for fresh waters used by fish and other aquatic life.

PHILOSOPHY OF STANDARD-SETTING PROCESS

Standards of water quality are, in general, similar to many other laws. Laws are simply rules of conduct that have evolved over the years. Laws are necessary to regulate the conduct of man—for his own good. If a law does not result in more good than harm (more "benefit" than "cost"), it relatively soon gets amended or even repealed. This same basic reasoning should and must apply to water-quality standards, or "laws."

Although some states have used water-quality standards for many years, water-quality standards on a countrywide basis obviously are new. Being new, it is inevitable that strong differences of opinion should exist on just exactly what the "law" should be, i.e., what limitations are needed so that the good resulting will exceed the harm resulting. Although water-quality standards are formulated by a relatively few people, both in the states and in the federal government, the standards must be set with the needs and desires of the public in mind. If not, the standards will be changed by the public, one way or another.

The basic question to be decided, then, by the law-makers is what limits should be set so that the public will receive the maximum net benefit from the law. In water-quality standards for temperature, limits set too high could result in a variety of harmful effects on aquatic life in areas near sources of heat, including, possibly, the death of some fish. Limits set too low will require existing and future thermal-power plants to install auxiliary cooling systems and pass the costs thereof on to essentially every person in the area, since practically all are certain to be power users. Protection should be provided to aquatic life—because the public wants, demands, and is entitled to it—but, on the other hand, the protection provided should not be so extreme as to penalize everyone in the area more than they, as a group, would be willing to pay, if they had a free choice, for the benefits provided aquatic life. Thus the law-makers are caught on the horns of a dilemma in trying to set desirable standards.

Actually, as everyone familiar with the situation will admit, the limits needed to protect, but not overprotect, aquatic life are unknown at this time. In such a situation, the tendency would probably be to set standards on the "safe" side. This might be popular with the public, because it is akin to being against sin. However, the public may not be aware how much the power rates will need to be increased to provide for the extra cooling facilities required.

In the writer's opinion, the best interests of the public are served if reasonable standards are set now, with intensive and extensive experimental studies being made to obtain the needed facts, and then the temporary standards can be adjusted, in line with the findings of the studies. Unless this procedure is followed, the "safe" standards are almost certain to result in the expenditure of tremendous sums of money—that could be saved by the public—for cooling towers, cooling ponds (lakes, actually), etc.

In line with this thinking, TVA is making plans for carrying on such experiments at the site of the Browns Ferry Nuclear-Power Plant on Wheeler Reservoir in north Alabama. You were informed of these plans this morning by Dr. E. B. Welch, in his discussion of Dr. Hawkes's paper.

MIXING ZONES

One of the provisions of most state-proposed criteria for control of stream temperatures involves mixing zones in the receiving body of water. Most of these criteria include such words as "after reasonable mixing, the temperature of the receiving stream shall not . . . ", "a reasonable and limited mixing zone may . . . "

In the writer's view, the mixing zone should be given more consideration.

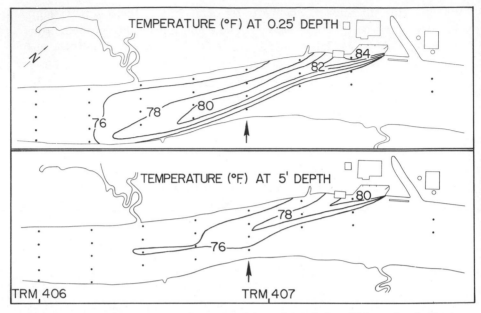

Fig. 1—. Temperature at surface and 5-foot depth below Widows Creek Plant, August 30, 1967.

Attention needs to be given to what is going on in the mixing zone and to the potential effects of the mixing zone on aquatic life in the receiving stream. Control criteria apply only outside the mixing zone.

A better understanding of just what is meant by a mixing zone is provided by study of Figures 1 through 4. The data shown were obtained below two TVA thermal-power plants on the Tennessee River in north Alabama. At the Widows Creek plant, on August 30, 1967, the volume of condenser-cooling water being discharged was 2,200 cubic feet per second and total stream-flow in the river was close to 47,500 cfs. Note from examination of Figures 1 and 2 that the temperature of the outflow from the plant was between 84° F. and 86° F., whereas the temperature of the river above the plant was between 74° F. and 76° F. The warm plume spread diagonally across the unstratified river (approximately 1,300 feet wide, here) in less than the upper 10 feet of water, and less than half the river depth.

What should be called the mixing zone here? The writer proposes that the thermal mixing zone be defined, in a physical sense, as limited to that heated portion of the cross-sectional area of the river or reservoir within which the isotherms are *not* essentially horizontal. At that downstream point where the isotherms become horizontal, mixing is not occurring. Thus, in

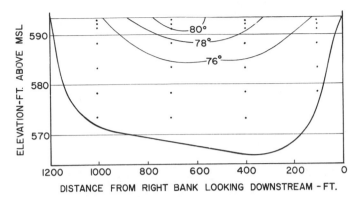

TENNESSEE   RIVER   MILE   406.9

Fig. 2—. Temperature distribution 2,400 feet below Widows
Creek Steam Plant, August 30, 1967.

Figure 2, the mixing zone at this location would be limited to the cross-sectional area confined within the 76° F. isotherm. Obviously, this represents only a small fraction of the total cross-sectional area. Thus, in this case, while the mixing zone extends entirely across the river in the surface strata, the mixing zone could not be said to constitute a thermal block to the upstream and downstream movement of fish.

At the Colbert plant, on September 8, 1967, the volume of condenser-cooling water being discharged was close to 1,900 cfs, and the total river-flow was approximately 55,000 cfs. Note from Figures 3 and 4 that the cooling water is discharged to a small creek, from which it enters the river on the left edge of a wide, shallow, over-bank section. Consequently, at the existing flows and temperature differentials, the cooling water flowed downstream in the shallow over-bank area along the left bank, mixing but very little with flow in the main channel. Here, the mixing zone, as defined above, extended from water surface to stream-bed in the over-bank area only, over a distance of approximately three miles. Here, also, no thermal block existed, but the basic characteristics of the mixing zones in the two cases are very different, one from the other.

At both of these power plants, the volume of cooling water is a small fraction of the total river-flow and, consequently, the mixing zone in each case is relatively small and creates no problem. However, in those cases where the thermal rise in the cooling water is, say, 15° F. or more, and the volume of cooling flow is perhaps a quarter or more of the total stream-flow, the mixing zone would be relatively large and surface temperatures in the zone might be objectionable. In such cases, provision should be made

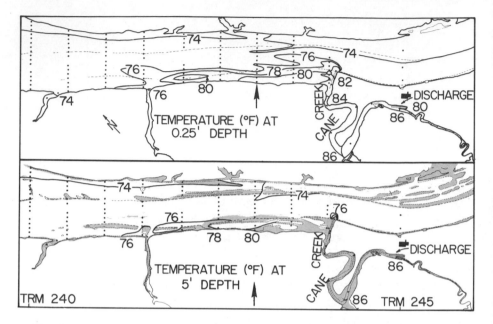

Fig. 3——. Temperature at surface and 5-foot depth below Colbert Steam Plant, September 8, 1967.

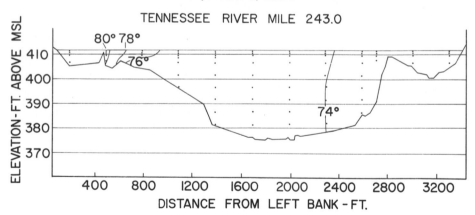

Fig. 4——. Temperature distribution 5,700 feet below Colbert Steam Plant, September 8, 1967.

for rapid mixing of the warmed water with at least most of the flow in the river. A multi-port diffuser on the bottom of the river—such as is being built at the Browns Ferry Nuclear Power Plant—will keep the mixing zone relatively small and thus avoid the creation of a large "hot spot" in the receiving stream or reservoir.

In the writer's opinion, for those cases where temperatures in the mixing zone exceed the thermal limits for the stream itself, the mixing zone, as defined above, should be restricted to not more than 75 percent of the area of the cross-section of the receiving stream. Fish and other aquatic life can thus pass freely upstream and downstream and, in addition, the flow bypassing the plant will contain adequate "seed" plankton to replace from new growth those plankton failing to survive the trip through the plant condensers.

I noticed that when he started to speak, Mr. Burd stressed the point in his thinking that artificial changes in water quality should be limited, such that optimum conditions are maintained or attained for aquatic life. This sounds good. It's like being against sin; it sounds fine.

But I think you have to look a little deeper into this thing; at least, I think you should. I think that water-quality standards should provide for optimum water quality. But in this case, *optimum* means that quality of water which will result in the maximum net benefit to the people in the region.

This is a broad definition, obviously. And under it, all users are considered together. I think it should be that way. Obviously, use for aquatic life is just one use. There are many others, and they must not be forgotten.

To obtain the optimum for aquatic life obviously requires a condition where some other uses cannot be optimum. In other words, aquatic life should be protected, no question about that. But the emphasis is that it should not be overprotected. In other words, protect it, but do not overprotect it.

The next point is that we really don't know how to separate those two. We don't know where protection stops and overprotection starts. Setting standards for thermal pollution, you might say, is similar to setting speed limits. In other words, the term *safe limits* was mentioned in Mr. Burd's talk. This also sounds good.

But as in safety regulations for highways, if we had only safety in mind, we would set speed limits that would be something like five or ten miles an hour. This obviously is not looking at the whole picture. You have to look beyond that. Safety is just one of the things that you consider. Travel time, highway travel, is very important. So naturally you end up with a compromise between safety and the other uses that highways have to serve.

It's the same way with thermal standards. A reasonable compromise must be reached between adding no heat and adding unlimited quantities of heat. There has to be a reasonable compromise reached to serve the best uses of all the people. Of course, the biologists naturally are somewhat biased; I guess you would say engineers are, too. But, after all, in natural resources, you have to think of all natural resources, not just one or two and not just one or two of the types of users.

You must have a whole group in mind. The objective must be the best way to serve the whole group—in other words, the public as a whole.

Getting back to this business of safe standards: as I said, it sounds good; but if we overprotect, as I think we are inclined to do sometimes, to get a safe limit, obviously somebody will have to pay for this.

Is setting these limits, on the safe side, really in the public interest? Would the public be willing to pay for these over-safe standards if they knew what those standards were costing? The answer is, they aren't told; they don't know.

We shouldn't, I think, administer standards that way. The public is entitled to know what's involved. This is a big subject. It is one that affects many, many people. I think the public is entitled to know the whole story. They have to be educated, obviously. That's part of our job here, to do just that, give them the whole story, not just a piece of it.

Obviously, the limits recommended by the Advisory Committee for aquatic life will require many cooling towers to be built at many existing power plants, and certainly many more in the future. These cooling towers will cost many, many millions of dollars. I am not saying they are unnecessary, you shouldn't do it. You certainly should do it. But let's be sure; let's not start off being really safe, over-safe in some cases, I am afraid. Let's find out. After all, the public has to pay for these costs, one way or another. The costs are passed on.

My suggestion is that we need to consider reasonable thermal standards until we know that they are not reasonable. That's the point of our proposals at Brown's Ferry, as was mentioned this morning.

As Mr. Burd has stated, we don't know an awful lot about this business. We need to find out. So let's establish some reasonable standards and then let's get real busy in a hurry finding out what we need to protect but not overprotect.

DISCUSSION/ Carlos M. Fetterolf and Loring F. Oeming

WE found Mr. Burd's exposition on the Federal Water Pollution Control Administration's past activities and current attitudes to be lucid and thorough. However, any relevance between our discussion and the points made in Mr. Burd's paper is pure coincidence, inasmuch as we prepared our comments prior to receiving a copy of his presentation this morning. As it so happens, many of our points are pertinent to his.

One point in Mr. Burd's presentation deserves direct comment, to clear the record. Mr. Bregman (1968), in his keynote address this morning, and Mr. Burd both said that some thirty-seven to forty states had received federal approval of their water-quality standards, to date. But neither mentioned how many of these approvals carry exception in the area of temperature criteria. For example, in the eight Great Lakes states, temperature criteria have been approved for only three. Of these, two do not conform to the United States Department of Interior's goals. These states will likely be asked to modify their criteria. This illustrates the "hang-up" the states are experiencing.

Those of us in attendance at this seminar who are concerned for the present and future conditions of our nation's waters can probably all agree that controls over thermal discharges are, and will continue to be, increasingly necessary or we wouldn't be here. The literature is replete with papers detailing the relationships of temperature and aquatic life. This literature, the speakers at the earlier symposium on biological considerations, and the speakers this morning have left little doubt that temperature alterations can have significant effects on the aquatic environment's suitability for the growth and propagation of aquatic life. Therefore, if man is to manage the aquatic environment to his advantage, he must control the artificial inputs which influence the ability of this resource to serve his needs best.

At this point, the agreement ends, and the disagreements begin. While the biologist is sure some controls are needed, we haven't heard any una-

85

nimity of opinion expressed as to what temperature limitations are necessary to prevent undesirable consequences or to insure a beneficial situation. The proposal of one biologist may be praised or criticized by others. Dr. Donald Mount (1968), Director of the Federal Water Pollution Control Administration's National Water Quality Laboratory at Duluth, stated at the earlier symposium that less is known about thermal pollution than the other water-quality parameters with which we're dealing. Carlos Fetterolf (1967; 1968), author of Michigan's intrastate and interstate water-quality criteria for aquatic life, stated publicly that the only criteria he lacked confidence in were those dealing with temperature. Secretary Udall was also dubious and, as a result, failed to approve our state's temperature recommendations. Disagreements such as this are stimulating to the individuals involved and it may be hoped that they will spark the needed research pleaded for at the earlier symposium. However, such disagreement is discouraging to enforcement agencies, cripples their programs, erodes public confidence, and is a serious impediment to those designing and financing future facilities which will have a thermal discharge.

APPLICATION PROBLEMS

Most of the disagreement so far has been over numbers: degrees of increase. How many degrees above ambient temperature in northwestern salmon streams; in midwestern smallmouth bass and walleye rivers; in southern bayous; and in northeastern estuaries? We feel these questions can be temporarily resolved from the existing literature of past research. However, this question should not be the *cause célèbre*. The important matters, both biologically and economically, are the mixing zone, the zone of passage, and the basis for control. It matters little whether the allowable temperature increase is X or $2X$ units. Of prime importance are the reasons for applying the restrictions and whether the mixing zone is 100, 500, 1,000, or 2,000 yards from the point of discharge, or whether the zone of passage must be one-third, one-half, two-thirds, or three-fourths of the river's width, and the reasons for specifications on locations of intakes and discharges.

Unlawful Pollution Defined by Statute

The water-pollution control agency must delineate restrictions on mixing zones and temperature changes. In Michigan, we have a statute (Act 245,

P.A. 1929, as amended from time to time) which, in Section 6, defines unlawful pollution, in part, as follows:

It shall be unlawful for any person directly or indirectly to discharge into the waters of the state any substance which is or may become injurious to the public health, safety, or welfare; or which is or may become injurious to do mestic, commercial, industrial, agricultural, recreational, or other uses which are being or may be made of such waters; or which is or may become injurious to livestock, wild animals, birds, fish, aquatic life, or plants or the growth or propagation thereof be prevented or injuriously affected; or whereby the value of fish and game is or may be destroyed or impaired.

Legal Interpretation

Nicholas V. Olds (1968), Assistant Attorney General for Michigan, has analyzed Section 10 of the Federal Water Pollution Control Act and advised:

Thus, the Water Resources Commission is vested with power to issue orders restraining discharges that may be made into the waters of the State so that they do not cause the injuries specified. It should be noted that any order so issued must be based on a finding of fact that the discharges unless restricted may cause any or all of the injuries specified.

. . . . . .

It is settled law in this state that riparian proprietors have a right to make reasonable use of the water which flows by their property, subject to those reasonable regulations which the legislature deems necessary for the public good.

. . . . . .

Thus, to deprive a riparian owner from discharging any substance into the waters of the State it must be shown that such a discharge is or may become injurious to one of the uses and values specified in Sec. 6(a). To require more, i.e., to prevent the discharge of an effluent which may cause no injury to legally cognizable public interests could be held to be a taking of property without just compensation. The sole purpose of pollution control laws is to so restrict and control the substances which may be discharged into the waters of the State, as to prevent injuries either to public or private use which may be made of the waters. Inasmuch as the Federal Constitution prohibits the taking of property without just compensation, the same principles would apply to the Federal Government in abating pollution of interstate or navigable waters as prescribed in Sec. 10 of the Federal Water Pollution Control Act.

. . . . . .

Since water quality standards are pre-determined restrictions on the effluents which may be discharged, they are subject to the same test of constitutionality. Thus, a standard of water quality which is unrelated to any injury that may be caused either to public or private uses and values or which would be an unreasonable exercise of the police power of the state, would be questionable under both the State and Federal Constitutions.

On-Site Restrictions

Unquestionably, thermal restrictions are a necessity. Their application is obviously facilitated by the establishment of water-quality standards. However, the various types of aquatic environments and the differing heat loads introduced appear to preclude the application of uniform restrictions. For example: a group of engineers and a biologist are currently considering the problem of disposal of heated discharges into Lake Michigan. One point discussed was whether thermal wastes should be discharged into water less than 30 feet deep. Another point was whether discharge points for thermal loads should be equipped with dispersion devices to insure mixing and conformance with temperature standards within 600 feet of the point of discharge. These questions were placed on the floor only as a starting point for discussion. Let's consider how these approaches would fit an existing situation in light of the foregoing.

The fossil-fuel James H. Campbell Plant of Consumers Power Company normally discharges between 600 and 900 cfs on the shore line of southeastern Lake Michigan (Figure 1). The plant has been on-line at partial power since 1962. The intake channel extends ¼ mile into Lake Michigan to a depth of 16 feet. Intake water temperature is raised 18° F. No cooling treatment is provided, except for the ⅝-mile-long discharge canal. On July 3, 1968, the discharge was about 600 cfs and the maximum temperature recorded was 83.5° F. at the discharge. An onshore breeze of 5 m.p.h. permitted puddling of the discharge. The ambient lake temperature of 65° F. was exceeded by 10° or more for a distance of ¾ mile and by 5° or more for about 1½ miles. At a depth of 2½ feet, the distance from the discharge, the water with a 10° increase could be found had been shortened to ¼ mile. The water increased 5° still extended 1½ miles. At a depth of 5 feet, the warmest temperature recorded was 67° F. in the immediate discharge area. Ambient temperature was not exceeded by more than 2°.

On another occasion this summer, an up-welling due to a wind shift occurred. The intake water was 55° F. Heavy surf prevented securing temperatures in the immediate vicinity of the discharge and the warmest area found was 65° F. This water became mixed within ½ mile up the coast and never extended offshore more than 150 yards.

This shore line area is very popular with resorters who either own or rent cottages or permanent homes. They seem to enjoy the warm water, and

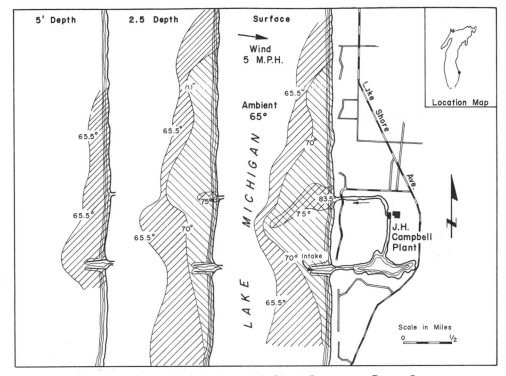

Fig. 1—. Thermal plume of J. H. Campbell Plant, Consumers Power Company, Port Sheldon, Michigan, July 3, 1968.

boaters and swimmers are attracted to this area. We have no record of complaints by swimmers of the heated discharge. The offshore beach area is shifting sand fully exposed to wave action. Benthic animals do not successfully colonize this type of wave-swept habitat along this shore line at depths of less than 10 feet. Shallow, sandy, exposed beaches, such as this, have historically been avoided by Michigan's fishermen, presumably because of the lack of a catchable fish population. No sport or commercial fisheries exist in areas of this nature, to date. Growths of the algae *Cladophora* are common on fixed substrates along this coastline, but observations disclose no increase in the warmed zone.

In summary, this appears to be an uncontrolled thermal discharge of appreciable volume which is not creating a biological or recreational problem. Without additional information to the contrary, it would appear unwarranted to require discharge into water 30 feet deep or to disperse the discharge to insure mixing to conform with standards 600 feet from the point of discharge. We learned today that the group discussing these points rejected them for universal application.

SITE SELECTION

Michigan's water-quality standards are prefaced by an original non-degradation statement which has been approved by Secretary Udall. It states that water with existing quality better than established standards will not be lowered unless and until it has been affirmatively demonstrated to the Michigan Water Resources Commission and the Secretary of the Interior that the change will not become injurious to any uses stipulated in Section 6 of our statute and that such lowering in quality will not be unreasonable and against public interest in view of the existing conditions in any waters of Michigan. It appears that it would not be difficult for a future thermal-discharger with effluent characteristics similar to the Campbell Plant's to claim his discharge would not constitute unlawful pollution.

However, if this same discharge occurred along a rocky shore on which smallmouth bass spawned or if the heated zone masked a small stream used by steelheads or coho salmon, a different attitude should prevail. Similarly, if the discharge was in an acceptable location from the biological viewpoint, but was of greatly increased magnitude, such as 2,000 cfs, with an increase of 28°, as has been proposed, we should be very anxious to learn more about what the damaging effects would be and what cooling techniques were planned.

Consultants Recommended

A third approach discussed by the group we mentioned earlier was that future plants be constructed in such a manner that initial or additional cooling devices could be added at a later date. This suggestion is not only practical, but opens a new avenue that has much to recommend it in the light of the present confused situation. We believe the enlightened control of thermal discharges offers unlimited opportunity for enforcement agencies and the discharger to negotiate equitably. The discharger and his engineers should have rational biological-limnological advice from consultants familiar with local problems, so that management is adequately informed and represented. The biological-limnological consultant should be used by the company in the earliest planning stages to advise on site selection, on intake and outfall placement and screening, on current studies, and on inventorying the pre-operational aquatic environment. This is a small investment for the avoidance of acrimonious debate or extended litigation.

Environmental Safety Factors

Thermal dischargers may feel aquatic and fishery biologists are a bit sensitive and defensive over heat loadings. This comes from past experience. Biologists have seen the aquatic environment degraded by one source after another. Each one started as "just a little bit which won't hurt anything" and then grew to unmanageable proportions. In the past, biologists have somehow been too straightforward or too naive to use the shields so popular in the engineering fraternity, safety factors and over-design. The time has come to acknowledge that our aquatic life and attendant values have a much more desperate need for safety factors than does a steel beam.

The control of thermal discharges is not an intolerable economic hardship, especially when viewed in conjunction with over-all pollution control. The attitude of doing as little as possible to protect our waters must give way to a philosophy that it is essential for a man today to assume the responsibility of insuring future generations their rightful legacy of clean water.

REFERENCES

Bregman, Jack I. 1968. "Keynote Address." Paper read at National Symposium on Thermal Pollution: Engineering and Economic Considerations, August 14–16, 1968, at Vanderbilt University, Nashville, Tennessee. (See pp. 3–14, this volume.)

Burd, Robert S. 1968. "Water-Quality Standards for Temperature." Paper read at National Symposium on Thermal Pollution: Engineering and Economic Considerations, August 14–16, 1968, at Vanderbilt University, Nashville, Tennessee. (See pp. 72–109, this volume.)

Fetterolf, Carlos M. 1967. "Water Quality for Fish, Wildlife, and Other Aquatic Life." In *Water Quality Standards for Michigan Interstate Waters*. Lansing, Michigan: Water Resources Commission, Department of Conservation.

Fetterolf, Carlos M. 1968. "Water Quality for Fish, Wildlife, and Other Aquatic Life." In *Water Quality Standards for Michigan Intrastate Waters*. Lansing, Michigan: Water Resources Commission, Department of Conservation.

Mount, Donald I. 1968. "Water Quality Temperature Requirements for Aquatic Organisms." Paper read at National Symposium on Thermal Pollution: Biological Considerations, June 3–5, 1968, in Portland, Oregon. (See pp. 140–147, *Biological Aspects of Thermal Pollution: Proceedings of the National Symposium on Thermal Pollution, Portland, Oregon, June 3–5, 1968,* edited by Peter A. Krenkel and Frank L. Parker. Nashville: Vanderbilt University Press, 1969.)

Olds, Nicholas V. 1968. "Memorandum from Michigan Department of Attorney General to Loring F. Oeming." Michigan Water Resources Commission, Department of Conservation.

DISCUSSION FROM THE FLOOR

Leon W. Weinberger: I think it would be appropriate to give Mr. Burd an opportunity to respond to some of the points made by the two discussers.

Robert S. Burd: First of all, Mr. Oeming mentioned a number of states that have had their temperature criteria excluded from approval by Secretary Udall. If it is any consolation, I would like you to know that the District of Columbia criteria were excluded from approval, also.

Mr. Churchill mentioned using the water resources to the maximum benefit of the people. This is his definition of optimum conditions, rather than marginal. He said aquatic life shouldn't exclude other uses. Of course, I think we can turn that around and say that other uses shouldn't exclude aquatic life.

He talked about not overprotecting aquatic life and making compromises. I think the problem that we are in today, with pollution throughout the country, is that we have been making compromises too frequently. My point is, compromises are not necessary, in that the technology exists for heat dissipation and the economics are feasible.

In reference to the question of whether the public would be willing to pay, I would suggest that perhaps engineers are out of touch with public opinion. I think public opinion certainly has changed in the last few years. I think the public would be willing to pay for adequate protection for aquatic life, given the facts.

I was interested in the discussion on mixing zones. This is something that I think needs considerable discussion. The suggestion has been made that a certain portion of the cross-section of the stream can be used as a mixing zone, as long as it is limited. I would be interested in some reaction from the floor as to whether there is some concern about how long a mixing zone we might have if we had one plant after another along the same bank of the stream, with the heated discharge hugging that bank of the stream.

I could visualize a situation where, perhaps, the cross-sectional area wouldn't be entirely affected by the heat of discharge, so that we might have many, many miles of the stream on one side dedicated to a mixing zone, and not available for aquatic life.

The question we face, in many parts of the country, is what percentage of an estuary or a lake we should set aside for mixing zone for a particular heated discharger. I know there was one proposal that said, well, only one or two percent of this bay will be set aside for a mixing zone for this one power plant.

What happens when the next power plant comes? Do they also get one or two percent of that bay? How about three, four, five, and ten power plants? What percentage of the bay should we retain for the optimum growth of aquatic life?

I think if we are to err, let's err on the side of safety.

R. S. Carter: I am a relative newcomer here. Perhaps you will forgive me for taking up your time today. Reference has been made to this mixing zone several times, and I have read it in the literature several times.

I get an indication that this is a good thing to do. Yet, when you consider point discharge into a river—I had one river in mind—it expands as a flattened, floating cone with natural and complete mixing, two-and-a-half miles downstream. At this point, the temperature of the river has been raised a couple of degrees.

During this time, the discharge water generally is near the surface, and therefore I would have presumed there would be an opportunity for more rapid heat dissipation. Also, while it is proceeding this two-and-a-half miles, I would think that the fish, having some degree of intelligence, would know how to avoid this cone.

If, on the other hand, you consider this mixing proposition, there should be a rapid drop in temperature of the discharge at the point of mixing, but with a subsequent very small loss as you proceed down the river.

I would like to ask you to comment on the relative merits of the two methods when contemplating locating plants as close together as possible on a river or lake.

To simplify my question: Have these things been considered, and is it true that you can lose more heat faster by mixing than by releasing the discharge at a point?

Burd: I might make a couple of comments but Mr. Oeming, as the administrator who has to face these problems, might better answer it. I sort of hesitate to sound like a biologist again, but correct me—all of you biologists—if I am wrong.

I am led to understand that there are some fish who feed at the surface, so that maybe the disposal of heated water near the surface in order to get the more rapid cooling possible, there, is perhaps not the best idea, depending on the species of fish present. Also, the fact that the oxygen absorption rate would be decreased, I think, is important.

I think the number one objective should be to keep the mixing zone as small as possible through the best possible engineering design of the outflow device. Perhaps you could respond, Mr. Oeming, about how it might be done in a practical situation.

Loring F. Oeming: I wish I could. I think I have a memory block here, because I am thinking about Lake Michigan, Huron, Superior, and Erie, which are special kinds of situations and not the kind that the gentleman is talking about.

While Mr. Burd was talking, I was trying to think of a case that might be used as some groundwork for this kind of an approach. The best I can come up with is a series of power plants on the St. Clair and Detroit Rivers, which are the connecting straits, you know, between Lake Huron and Lake Erie. There is there a series of power plants, but nothing of the magnitude that we are now talking about of nuclear-power installations on Lake Michigan.

We have had some benefits from this spreading out, and it hasn't been done to achieve a mixing zone. But there is good mixing in the Detroit and St. Clair Rivers. I guess the benefits, so far, are that the river is open longer in the winter, and everybody is yelling these days about extending the navigation season so we don't have to put salt in the river up there to keep it open for the boats. It will be kept open by warm water.

Milo A. Churchill: I would like to speak to that point. We are building a multi-port diffuser at our Browns Ferry Plant on the Tennessee River near Decatur, Alabama. This diffuser consists of large pipes—in this case, there are three of them, because there were three reactors. On the ends of each of these pipes there is a 600-foot section in which there are many small holes.

Anyway, Don Harleman and the hydraulic people in the laboratory at M.I.T. and Rex Elder and his group at Norris checked various designs and concluded that many small holes, two inches in diameter, something like 700 on the end of each of these large-diameter pipes, having jet velocities in the order of 7000 feet per second in pipes measuring 17 feet up to 20 feet in diameter with a total discharge of 4400 second-feet, initiates rapid diffusion and results in very quick mixing in a small mixing zone. Obviously, if we didn't use that approach, we would think about discharging this warm water on the surface. We would obviously lose heat much faster if we discharged at the surface, but at the expense of a lot of aquatic life in the surface layers of this reservoir. So, rather than sacrifice the upper layers, we decided that the best approach in this case was to use the diffuser as a promoter of rapid mixing, to get the temperature down to a relatively small temperature increment, such that there would not be a large hot-spot out in the lake.

G. Earl Harbeck, Jr.: I wish to avoid getting into the effect of thermal pollution upon aquatic life, as this aspect is not in my field of knowledge. I must emphasize that the most efficient way of disposing of this heat is to spread it on the lake or river in a thin sheet. Almost all of the heat that is being added by the plant is being disposed of by three processes, as follows: (1) an increase in back-radiation from the water surface, (2) an increase in the energy utilized for evaporation, and (3) an increase in the amount of energy conducted from the water surface to the atmosphere. These are all surface processes, and if you want to dissipate the maximum amount of heat in the shortest distance possible, you will put that hot water in a thin layer on the surface.

R. E. Nakatani: I can make some comments about the behavior of salmon in warm water. We have worked a little with other species, but those who attended the meeting at Portland will recall that John S. Alabaster provided some interesting information about the behavior of other species (trout, roach, perch, bream) in temperature gradients.

In controlled laboratory experiments, it is well demonstrated that fish will select a particular temperature in gradients set up experimentally. The selected temperature or the preferred temperature will be higher than the past temperature of acclimatization if this temperature was low. If, on the other hand, the temperature of acclimation was high, the selected temperature will be lower. This is one type of behavior you might observe.

In the field situation at Hanford, the problems the salmon faces, migrating past Hanford reactor plumes, can be considered of two kinds.

One, the young seaward migrant, with ability of swimming against only one-foot-per-second velocity, essentially gets carried along in fast waters of about 5-to-6-feet-per-second near reactor plumes. The volitional behavior or avoidance behavior of these young fish doesn't mean very much in fast waters. Fortunately, there is evidence that most of the juvenile migrants hug the banks, with lesser numbers in mid-channel. This distribution near the banks is also true for the upstream adults.

The problem the upstream adult faces is different, for one reason, simply because the adults are swimming upstream, and if they enter a plume, they are subject to increasing temperature gradients. They have considerably greater swimming ability and the avoidance behavior of these adults becomes very important for survival.

With reference to the question of what the temperature pattern of the discharge should be, I believe more understanding of fish behavior in relation to warm waters is necessary. From a very pragmatic viewpoint, we need to keep in mind the important species, the so-called desirable species we are trying to protect. Clearly, there is a wide biological variation in behavior, and what is satisfactory protection for the salmon may not be satisfactory for other species. Each site with its own special ecosystem will require its own study.

Carlos M. Fetterolf: The question was how bright fish are, in avoiding heated-water discharges. I think that this relates to their age; whether they are going downstream or upstream; what time of year it is; what the ambient temperature is; and many other things. Their response to temperature inputs is very variable. Sometimes, they will orient to the temperature inputs. At other times, they will avoid them. Much depends on the difference between the input and the ambient temperature and also on what the fish has in mind.

Nakatani: Mr. Fetterolf has made a good summary. In addition, I wish to stress the complexity of fish behavior. Brett has determined that the preferred temperatures for young salmonids extends from 12° to 14° C., and fish generally avoid temperatures above 15° C. I don't think much work has been done on how a school of young salmon reacts to temperature gradients. It is interesting that, if a school of fish are placed in an apparatus with an electrical gradient, it turns out that the school turns and stops penetrating the higher electrical field when the lead fish turns. In other words, the behavior of the school is highly dependent on the behavior of the lead fish.

In nature, salmon biologists have long known that young sockeye salmon resident in a lake are generally found at the thermocline, indicating that fish seek out some preferred temperatures.

Sam Posner: I am a bit surprised at the nature of the discussions at this symposium. I get the feeling that the trend of thought has been transferred from the control of thermal pollution to the development of better ways to mask it. Is this meeting not designed to recommend ideas for the federal government to set criteria for the future? If those of us interested in the control of thermal pollution are not adamant about the basic criteria set down at this time, we will develop the same problems that the air-pollution control agencies are involved in today.

The amount of conventional and nuclear-power plants projected earlier by our colleagues from the Federal Power Commission should not only dramatize the magnitude of the thermal pollution with which we will contend in the next 15 to 20 years, but should also provide the impetus to design a compact, all-encompassing program.

Norman Brooks: From the hydraulic point of view, it is possible to design a variety of different ways to discharge water, such as thoroughly mixing it or skimming it on the surface. But it is not altogether clear which method is most desirable in different cases.

In fact, I think we ought also to look at the energy cycle for a whole power-plant system, including its environment. This is analogous to recent efforts to think about waste management in a broader sense with air, land, and water pollution considered all together.

For example, if we regard thermal pollution as being only a *water-*pollution problem, we are making a mistake; for instance, the rapid rejection of waste heat out of the water into the atmosphere may simply transfer a thermal-pollution problem to the atmosphere.

When we generate electricity, we produce waste heat. The *ultimate* disposal of this waste heat can only be by long-wave radiation into space. It may go, first, into the water-stream and then into the atmosphere, but ultimately it will stay on earth until it leaves by long-wave radiation. That's the only route it can leave by. In fact, the energy of the fuel came, in the first place, in the form of solar energy which was fixed by photosynthesis, and then finally it goes back again to space.

Also, the consumption of electricity produces waste heat. This usually doesn't get into our water-stream, but I think many of you know that we have a gradual warming of the urban surroundings. I think much of this warming is associated with the consumption of electricity, as well as the burning of fuel in the cities.

Thus, this waste-heat problem is a lot more comprehensive, than, say, just keeping it out of the water bodies. In fact, in the very long run, the next hundred years, say, we may have to be more careful about how much energy we recklessly consume altogether on Earth. As Revelle has said, Earth is like one big space ship; for steady state, there must be an excess of radiation over incoming radiation to get rid of man's production of energy.

The only way that's going to happen is to have a gradual warming of the earth. That might seem rather far-out; but in the metropolitan Los Angeles area, the amount of fuel burned, electricity consumed, and waste heat produced by power plants already amounts to about three percent of the incoming solar radiation.

If this energy consumption doubles many more times, it will become large compared to the solar isolation, and there will certainly begin to be impact on the over-all geophysical heat balance.

Now, I think there is still one other aspect of heat that deserves some attention. The other kinds of pollutants are susceptible to treatment processes, usually involving doing some work, or pipeline transport, also requiring energy.

Any time energy is used for pollution abatement, however, more power must be generated somewhere else, and that makes more waste heat. Thus, if you think of "treating" waste heat by concentrating it and transporting it somewhere, in reality all you are doing is running a refrigeration system, which takes work; this means more power and more waste heat.

So there is no way, really, of developing a basic "treatment" process for waste heat, other than radiation.

Richard Foster: I have a short comment on the general subject of the interaction of water-quality criteria for fish versus optimum use on a multi-purpose basis. I think we have heard quite a few comments relative to whether one should set temperature standards based on criteria for the optimal range for desirable species, or on criteria that represent the tolerable range. The question involves the use of the standards in respect to the total optimization of all water uses.

First, I would like to point out that I think we are going too far if we expect water-quality standards to accomplish this purpose, because water-quality standards, by themselves, should not be used to make a socio-economic balance required for optimal use.

Water-quality criteria for fish should be and are being selected by aquatic biologists. I think it is quite legitimate for these biologists to try to choose numbers which are in the best interests of the fish. The criteria so selected should not, however, be viewed as ones which are intended for total optimization. I think somewhere along the line we have to introduce an additional step in the process. This additional step is in the form of zoning.

We have talked about zones of mixing, but we also have to talk in terms of zoning of the streams themselves. We have to decide whether the principal use of a stream is going to be for fish of a particular kind and solely for fish use, or whether that stream is to be used for multiple purposes.

At this stage of the game, it is desirable to try to keep water-quality standards as simple and as straightforward as possible. To expect that standards are going to optimize water use is going a little too far.

Larry B. Vaughan: There has been substantial discussion here today, concerning water-quality criteria. I would like to ask Mr. Burd about the other half of the water-quality standards—that is, the implementing portion.

I have read recently that the FWPCA has had some poor experiences in enforcing their standards. I am wondering if you would give us the benefit of your experience as to how this implementation program is proceeding.

Burd: First of all, we consider the implementation plan a very important part of the water-quality standards, in that they do tell us how these criteria are going to be achieved.

You say we have had poor experience with enforcing them; but we really haven't had the experience, yet. The standards are too new. I would like to say, first of all, that we expect the states to be administering the water-quality standards, and we would look to the states, first, to enforce the standards, and only if there is a failure on the part of the state to enforce the standards would the federal government become involved.

Frankly, we anticipate that the states will meet their responsibilities. They are standing behind their standards and will be enforcing them themselves. But at this point, a year from the time the standards were adopted, it is premature to make a judgment as to how they are going to work out.

However, I am sure that they will work out. I certainly hope that they do, because if standards don't work, I think the alternative the Congress may think of next would, perhaps, be less acceptable to the states, municipalities, and industries.

I would like to respond to the earlier comment. I don't feel that the protection of aquatic life and industrial development are necessarily incompatible. I think an example of this is the non-degradation policy that the Secretary announced and which Mr. Oeming referred to. This policy applies to high-quality waters, where the water quality exceeds that of the established standards.

Because there was a definite lack of information, the Secretary felt it necessary to develop a policy to protect high-quality waters to make up for the imperfections in the standard-setting process.

The non-degradation policy anticipates that the necessary socio-economic factors would be considered, and perhaps some water-quality deterioration will have to be allowed; but it would be limited, because the best practicable treatment or control would be provided.

I think this policy is consistent with the states' approach to this problem in the past and the approach that they will follow in the future. So, in terms of thermal pollution, I think what it means is that if that mixing zone in high-quality waters can be a quarter of a mile long, rather than two-and-a-half miles long, by the installation of heat-dissipation equipment, and assuming the technology and the economics are feasible, we would have it. I think we should generally be looking for the best practicable control of the heat discharged, unless there is absolutely no effect on aquatic life, propagation or other legitimate uses.

This can be demonstrated, I am sure, in many coastal water-discharge situations; not discharges to estuaries, but discharges to the open ocean.

Fetterolf: There has been concern expressed from those in the audience that the respondents this afternoon—and perhaps even the biologists speaking this morning—don't seem especially concerned about thermal pollution. Such is not the case. We are very much concerned about thermal pollution. However, I am a biologist working for a control commission. On that commission is the Director of Conservation, the Director of Agriculture, the Director of Highways, the Director of Public Health, a man representing municipalities, a man representing industry, and a man representing sportsmen's groups.

It is my job to operate within a legal framework that is provided to me and to our commission by the State of Michigan by statute. Therefore, I cannot limit my deliberations to what would be the optimum conditions for aquatic life, because my Commission represents a multiplicity of interests in the management of water.

The biologist who is ultimately concerned with minute changes in aquatic life is usually a research scientist or is associated with an academic institution, or is interested in fish management only. He can demand that all heat be dissipated before it is introduced to a mixing zone if he feels this will result in optimum conditions.

So there is a difference of opinion among biologists according to who they are employed by and what their job is.

Charles Waselkow: With respect to the actual enforcement of the various standards, what are the nature of the penalties that are set up now? Historically, violations were about $100 per day. Is there a great variation among states?

Burd: Mr. Oeming can answer that question better than I, but there is a great variety in the penalties incorporated into the various state statutes. Some of these penalties aren't very large. But I think there is a next, more formidable, step that the state can take.

For example, if someone chooses to ignore a small penalty, I think an injunction can be secured and some correction of the particular problem assured in that way, if the monetary-penalty approach doesn't work.

Oeming: Most of the states I know about—and I know about few of them —have a penalty provision. It does vary. It varies upwards up to $1000-a-day penalty, for a violation of the statute itself or a violation of an order or rule of the administrative tribunal.

This is one course, certainly, you can follow, but I wonder how much good it actually does? It is going to apply to many states, but how much does it do to get the problem solved? All the statutes that I have seen provide for an enforcement procedure parallel to this. You don't have to go via the criminal route. You can go via the civil route. Here, you go to court for enforcement of the order. Once you go to court for enforcement of the order or a regulation or rule, and the court affirms that order, then you have got a real tie to get this job done.

There have been many arguments, particularly from conservation groups: why don't you fine people? Well, yes, you can fine them, but that isn't the answer. In fact, when you come down to practicalities, gentlemen, $500 a day penalty—there were a couple of cases I was involved in, where it would have been cheaper to pay $500 a day penalty than to provide pollution control or correction of the pollution problem.

What do you want? Do you want penalties, or do you want the problem corrected? The routes are there. Both procedures are there to do this, in every state that I know of, and I know about most of them.

George Rand: I don't profess to be a biologist. I am not a sanitary engineer. All I ask is how best we should live up to the laws of the states we operate in?

One of the things I had hoped to learn this afternoon, Mr. Burd, is what type of standards the federal government actually is driving for, on temperature. I believe we have heard, from another speaker, that in the Great Lakes area only one state had satisfied the federal government. We hear of all these standards being approved, with exceptions.

My company operates all over the United States, and it disturbs us when we hear of these exceptions and are uncertain what the federal government is driving at.

Possibly I missed the general point this afternoon, but as yet I haven't heard what you really want.

Burd: I think in the body of my talk I mentioned a couple of general guidelines. One, the standards that are established shouldn't deviate too greatly from the existing water quality; and two, they shouldn't deviate too greatly from the National Technical Advisory Committee recommendations.

To be more specific, we are looking for a minimum of two numerical parameters or criteria for temperature. One is a maximum-temperature limitation. This maximum, consistent with existing water quality, doesn't mean that the maximum has to be pegged right at the temperature maximum that exists today in the unpolluted stream, but it shouldn't vary too greatly from that, in terms of a few degrees. And, ideally, this maximum should be tailored to different streams.

Many of the states adopted a uniform maximum of 93° F. throughout the entire state, when perhaps they have streams in the hilly or mountainous areas of the state where the maximum never gets above 80. We feel the divergence there is too great, and the maximum for the cooler streams should be closer to what the existing quality is.

The second general guideline, once again, is some consistency with the National Technical Advisory Committee recommendations. This doesn't mean the standard has to coincide word for word with the recommendations, because, within the body of that report, particularly in the

Aquatic Life section, the Committee points out that these are general recommendations and have to be considered in light of local conditions. But the divergence from their recommendations should not be too great.

In addition to the maximum numerical limit, we feel it is necessary to adopt a limit on the change from the natural background temperatures. A suggested value is the five-degree-Fahrenheit limit recommended in the NTAC report. This is not five degrees just for one discharge; it is a cumulative five degrees. The value that is adopted by any particular state doesn't necessarily have to be precisely five, but it has to be consistent with that recommendation.

The way this would be administered would be to go upstream from a major metropolitan area, like Nashville, determine what the natural background temperatures are, and add on five degrees to that value and expect the downstream area to comply with a maximum that is within this five-degree limit.

So those are the two minimum requirements that we are looking for: a maximum limit and a change from background limit. And we are looking to see how consistent the state standards are with existing temperatures and with the National Technical Advisory Committee report.

S. Leary Jones: We are mixed up in this temperature mess, also. Mr. Udall has been extremely critical of the states—about six, I guess—that have had temperature standards adopted. As he says, there is something wrong with the other forty-four.

He is right. But we can't find out what it is. I mean this seriously. The governor has written Mr. Udall three times and asked him what temperature standard is wanted. Each time, the letter has come back, saying they are very glad that the Board is willing to consider another standard.

You mentioned the five degrees. We know what the ultimatum is from Washington: all standards must have a five-degree rise above the ambient temperature which will be the upstream temperature. Why won't Mr. Udall put that in writing? Seriously?

We are wasting a lot of time in trying to reach a temperature standard. I don't think the fish worry too much about whether we have five or ten.

We have had almost no fish killed in Tennessee from temperature—and I say "no," because I can't think of any, outside of small ditches.

There are about 600,000 organic chemicals, and we have only three specifically listed in our standards. We are getting far more fish killed from other items than temperature. But temperature has become a national issue.

The governor finally decided, after the last letter, that it was just a waste of time. But he asked twice for the FWPCA to send a responsible person down to meet with us and tell us what temperature they want.

Take the word back and get Mr. Udall to put out a statement that no state standard will be approved unless they adopt five degrees.

Burd: I would first like to say that the adoption and approval of standards is a negotiated process. One reason why Secretary Udall hasn't sent out a letter saying you have to adopt five is that this is not the case.

I think, if you look at the temperature standards that have been approved, you will notice a significant variety. And they are not all the precise recommendations of the National Technical Advisory Committee report. For example, a couple of states have chosen to take the approach that Dr. Bregman mentioned that Vermont Yankee was proposing: that is, having a sliding scale, where an allowable temperature change greater than five degrees was prescribed for the cooler months, with something less than five for the summer months.

So the Secretary is not telling you what Tennessee or any other state has to establish, because he feels this should be determined, to some extent, on the local conditions. I think probably what is required here is more communication between your biologists and the federal government's biologists.

I find myself very uncomfortable, as an engineer, in the middle, trying to referee various points of view expressed by biologists. But I think we can resolve this matter with Tennessee and other states if you present the scientific data you have pertaining to your state and your streams and your fish species and let the federal fisheries people respond to it. Then I think we can all work this out and come up with a satisfactory temperature standard.

Jones: I will keep this short, because this certainly isn't going to settle anything. These famous standards in the Green Book, the national recommendations, were published—when? A month ago?

Burd: The recommendations are not too different from the Interim Report that was released on June 30, 1967. However, there are some relatively minor changes in the aquatic-life recommendations.

Jones: When were the standards submitted to the Secretary? Prior to June 30, 1967?

Burd: That is correct.

Jones: And my kid almost whipped me once when I suggested he guess what I had in my pocket and if he didn't I was going to whip him. Seriously, Mr. Burd, we have met with the representatives from your office. The tape recorder was sitting on the table. We know exactly what their feeling is. There is no basis for picking five or ten degrees except by flipping a coin, except in specific cases.

But on a general basis, no. We also must meet the standards of the surrounding eight states. We have. Can that many people be wrong? I don't think the standard is worth a damn, because I don't think the fish can read.

Burd: The Interim Report did recommend five degrees. However, the maximum-temperature recommendations were changed; and in the final report, the maximum temperature was based on the species of fish. You and I both know that something other than five has been approved in a state standard and that this would be possible in your state, too.

But, recognizing that we have a lack of information, I think we are obligated to consider seriously the recommendations of the National Technical Advisory Committee. That group was comprised of well-informed fisheries specialists from around the country, and in lieu of other information, I think we have to give serious consideration to their recommendations.

Perhaps you in the state of Tennessee and your biologists and the university biologists do have some scientific data concerning your particular conditions and the fish that live in your waters that would justify some other number. I think we need to take a look at that scientific data.

Fetterolf: I feel I would be letting every biologist in the room down if I did not make some comments, despite the fact that Mr. Oeming says, "Stay out of this." It makes a lot of difference to the fish whether we are talking about five degrees or ten degrees. However, hot water doesn't often produce a fish kill, so you can't evaluate the effects of a thermal discharge on the numbers of dead fish that you find. What biologists are trying to do is provide satisfactory living conditions for the growth and reproduction of fish and other aquatic life. As I said before, it is important to the fish whether we allow a five- or ten-degree increase, but of ultimate importance is the size of the mixing zone. Don't forget that. The numbers aren't as important as the size of the zone.

Wright: For some two-and-a-half hours, now, we have talked about standards and mixing zones and, as near as I can tell, there has not been anything in this discourse relative to lakes. It has all involved streams or, at the very best, open-ocean discharges—with the exception of Lake Michigan, of course.

It seems that the general consensus is: Let's not raise it more than five degrees, and let it go downstream, or put it through a pipe with holes and mix it up, and to hell with it; it has gone away.

If we deal with lakes, it doesn't go away. We can put it on the surface. It will sit there, and it will keep the lake from mixing, and this is not necessarily a good thing. Ultimately, we can mix it in the lake, and it is going to stay there until it finds some way to get out. If the turnover time is ten or fifteen years, it will stay. Some, of course, will be lost to radiation. Mr. Brooks commented that three percent of Los Angeles's total insolation now is derived from electrical generation in Los Angeles. We are going to have cases where thirty percent of the heat in a lake is dumped in from thermal sources, other than solar insolation. What is going to happen is that you can't mix it away and it is not going to be very good to put it on the surface.

I wonder if the engineers have given any consideration to a concept involving heat units per water-volume in discharges? We have a big lake; maybe we can put a medium-sized plant on it. But for God's sake, let's not put a big plant on a little, tiny lake, even though we can perhaps stay within this five-degree range or whatever the figure is going to be at a given time.

Mr. Burd, would you want to comment on this concept of heat-volume per lake-volume for something other than streams where the heat is quickly mixed with flowing water?

Burd: I agree that if the lake is very small, not a Lake Michigan, we should minimize the input of heat. Temperature standards for the lake should be similar to that for a stream supporting the same type of fish. The National Technical Advisory Committee goes on to recommend that, if the lake is stratified, the hypolimnion should be protected by any temperature increase. The epilimnion should be protected by a lesser allowable temperature increase than, say, the five-degree increase that is recommended for waters where we don't have stratification. I think we should look very carefully at proposals to discharge heat into small bodies of water.

Allen Agnew: I am a geologist, so I can't take sides in this biology dichotomy. I do represent some sixty universities which make up the Universities Council on Water Resources. Most of you, I imagine, are alumni of these universities.

You recognize that we are attempting to educate and train the people who will be operating within your state or federal regulatory agencies, as well as the personnel who are employed by your companies.

These universities have interdisciplinary programs, and they include not only the sciences and the engineering aspects of water problems, but also the social science aspects. This is a point that I want to make. We have heard references to the decision-making process here this afternoon, and I think that much of what we are arguing about is mainly just that: Who makes the decision, and what institutional arrangements are going to provide the framework for the decisions.

Gentlemen, this is an educational process, and we in the universities are attempting to produce a product that will give us some of the answers in a few years. Of course, that doesn't solve your problem right now, for tomorrow night is already with us.

I am one of these people who, I suppose, like Mr. Churchill, and, I think, like Mr. Oeming, here, feel that there still is some esthetic beauty in a neat engineering structure, such as a properly designed cooling tower. I was horrified this morning to learn that such a tower is "an eyesore," a "piece of esthetic pollution."

The point that I would like to get across here is, of course, that we can pay for as much regulation and as tight restrictions and criteria as we wish. We have the money and we can do it. However, I take violent exception to the statement that I have heard two or three times today: that the public is aware of all facets of the problem, and the alternatives —because the public is not completely aware. The public needs to be educated. And, furthermore, I think that there is a limit to the amount that we can lead the public along by the nose and make the public's decision for it. I think that we need to make sure that the public gets all the facts.

Furthermore, can we, as individual biologists, be sure that we are speaking the same language to each other when we take roughly the same stretch of a stream and come up with different interpretations from the same physical data?

I think that a give-and-take exchange such as we have been seeing here today is very helpful. But we should not expect that out of this exchange is going to come some rigid set of national specifications that should regulate (should over-regulate) us, because I would far rather look for reasonable standards than for overstrict and inflexible ones.

William L. Klein: Has there been any discussion or decision within the Department of the Interior concerning heated discharges from municipalities? The ORSANCO staff has been looking into the Ohio Valley situation and believe there are a number of municipalities, as well as industries, which may require cooling towers.

The difficulties encountered in planning cooling towers are numerous, not the least of which is the public relations involved in convincing the cities of the necessity for cooling towers. This is particularly true if the cooling tower is needed in the winter months.

Burd: This has come up occasionally in the deliberations around the Department about standards. I particularly remember one state standard where they excluded municipal-waste discharges from the application of temperature criteria.

We questioned this, even though we assumed that putting a cooling tower at the end of a sewage-treatment plant is unreasonable. This state, by the way, did modify its standards and does not grant an exclusion to municipal-waste discharge, but they don't plan on installing cooling towers, either. I think the waste is going to be discharged in some other direction.

This point has been raised by states, particularly with ephemeral streams where, in the wintertime, when maybe half the flow of the stream is domestic sewage, the temperature in the stream may be raised some 15 degrees over the background temperature.

We, as everyone knows by now, have a recommendation of five degrees over background. Certainly we are realistic enough to recognize that there are these low-flow situations and they should be accounted for in the particular standards and the 15-degree F. increase would not be considered unreasonable, due to the source of the heat input.

Heat is also discharged by a number of industrial facilities, such as pulp and paper mills, and there can be a significant increase in temperature due to irrigation-return flows. We do look at these heat sources a little bit differently from the way we look at the heat input from a power plant. We put them in a nonconventional category and admit there is not an awful lot that can be done about them.

Leon W. Weinberger: I know there have been some grand proposals, as far as industry is concerned, to make use of cooling towers and biological treatment units, as well. Some of these, of course, would make some sense and not cause too much of a problem.

I think the approach, however, has not been so much from the point of view of thermal-pollution control, but rather from a point of view of the destruction of organic material in the use of the tower.

Keith Fry: I work for the International Paper Company. Mr. Hawkes this morning made mention of the fact that temperature increases coupled with organic loads in a stream presented quite a different problem from a pure temperature load in terms of the aquatic environment that the biologists speak of.

Mr. Burd's last comment, to the effect that they look at temperature differentiations in a river from a power-plant source quite differently from the way they look at a temperature differential caused by a discharge of a sanitary-sewage-disposal plant or perhaps any other kind of waste-disposal plant. This seems to be a little bit inconsistent, with respect to the demand to protect the water from the temperature increases, if, in fact, the presence of organic material tends to accentuate the problem of the increase. I would like to hear some comment, either from Mr. Burd or from Mr. Oeming.

Burd: I will just make one comment. The Federal Act talks about technical and economic feasibility, and I think this is a factor here. I believe the federal government is accused sometimes of being too unreasonable in its requirements, but I don't believe this, and there is that constraint in the Act—technical and economic feasibility. I think that probably comes into play in this situation.

THE COOLING OF RIVERSIDE THERMAL-POWER PLANTS

THE yield coefficient of thermal-power plants is quite low. In the most modern units burning conventional fuels, only 40 percent of the heat content of the fuel is transformed into electricity. Almost 50 percent has to be returned to the "cold source." The position with respect to nuclear-power plants is less favorable, still.

The continuous and rapid increase in the size and power of plants makes it necessary to dissipate considerable quantities of heat. In the case of a plant of two million kilowatts, for example, two million tons of coal equivalent of returned heat have to be dealt with.

The cheapest way of dissipating this heat, in terms of both investment expenditures and the over-all efficiency of the installation (and therefore its operating costs) is to evacuate it through the water of a river, lake, or estuary, or the sea.

This technique has been in use for a long time; but in the past, no particularly serious problems were posed by the slight increase in the temperature of a body of water of limited volume. Now, however, it is often essential to exploit the use of surface waters to the maximum, given the limits set by the existence of other water users. Correspondingly, it has become necessary to develop the most accurate knowledge possible of the thermal mechanisms brought into play when heated water is evacuated.

Different problems have to be faced in different circumstances. Considering riverside-cooled plants alone, the following cases are far from exhaustive:

(a) *Plants cooled by water drawn from non-navigable rivers.*—The structure providing water intake (wharf, dam) definitively separates the headrace from the tailrace junctions with the river. The only problem to be resolved is that of dissipating heat along the length of a fairly turbulent stream in order, for example, to determine at what distance downstream it is feasible to install another power plant, or some other installation for which fairly cold water is required, such as a filtering plant to supply drinking water.

(b) *Power stations situated on the banks of rivers which have been rendered navigable by appropriate navigational structures (locks, dams).*— The problem is very similar, but here the current moves only slowly, and the surface of the water is relatively unbroken. In some respects, its characteristics are the same as lake cooling, but the movement of vessels limits thermal stratification. Apart from the question of downstream cooling, there is the further problem (which may be dealt with separately) of recirculation, i.e., warm-water currents which may form between intake and reflux points during periods of low rates of flow.

(c) *Plants located some distance from a river, whose used waters exit into a borrow-pit or a pool which communicates with the river.*—Over and above the problem of the cooling process which occurs along the length of a fairly cold river (i.e., the mingling of the warm water vented from the plant and the river's cool water), it is necessary to examine the cooling of a body of water which undergoes a constant rate of temperature increase throughout the year.

(d) *Power plants connected by long headrace and tailrace canals to a river whose surface level is maintained practically horizontal (at least during periods of limited flow) by a dam.*—Where the river's rate of flow exceeds the rate at which water is pumped out of the plant, the problem is the standard one. But where the contrary is the case, recirculation occurs, and there is need to consider both the cooling of the river downstream from the dam and the dissipation of heat by the water surface formed by the headrace and tailrace and the section of river between them.

(e) *Power plants installed on estuaries in which the direction of the current reverses with each tide.*—This is the most complex case. Generally, rather than to try and assess cooling downstream from the plant, an attempt is made to determine the maximum increase in temperature at the cold-water intake point. In some respects, the problem is related to that met in the case of lake-side generating stations (mixture of warm and cold currents). In other ways, it resembles that of run-of-river plant cooling (cooling a body of water which has first been carried away from the plant by the current and then brought back); study of the speed of the currents is essential.

This last case will have to be dealt with in France within a few years, when large power plants may be expected to be installed on the banks of estuaries.

At present, Electricite de France and the public authorities are attempting experimentally to develop methods for understanding the cooling mechanisms and the mingling of warm and cold waters by observing the conditions ruling in the case of a number of existing power plants.

The results obtained for the plants at Montereau, Vaires-sur-Marne, and Beautor, all located on the River Seine or its tributaries, are presented below.

Description of the Plants Studied

It should be specified at the outset that, in all modern French installations, all circulating pumps are driven at one speed only, so that their rate of through-put is constant (although the Vaires-sur-Marne installation is an exception, in that its pumps can be operated at either of two speeds). The rate at which the water is heated is therefore approximately proportional to the power output of the turbogenerator, reaching 7° to 8° Centigrade when the generator is yielding maximum output.

The Montereau power station

This plant falls under category (b) above, as it is located on the Seine, which is a navigable river. At low water, the current moves very slowly (about 10 centimeters per second). The disposition of the installation is such that no hot-water recirculation phenomenon have ever been observed. Details of the plant:

Two 125,000-kilowatt sets and two 250,000-kilowatt sets (maximum intake rate: 29 cubic meters/sec)

Located on the banks of the Seine, some 90 kilometers upstream from Paris

The rate of flow of the river in severe low-water conditions is 25 cubic meters/sec, increased to 50 cubic meters/sec by reservoir dams constructed upstream.

The Vaires-sur-Marne Power Station

This facility also falls under category (b), but recirculation is a fairly frequent occurrence. Details:

Two 250,000-kilowatt sets (maximum rate of water intake, 19 cubic meters/sec)

The plant is located on the banks of the Marne, approximately 30 kilometers upstream from Paris.

Rate of flow in severe low-water conditions: 11 cubic meters per second. (In four years' time, this will be increased to 40 cubic meters/sec by a reservoir dam to be built upriver.)

The river is rendered navigable in the neighborhood of the plant by a dam sited 2.9 kilometers downstream. The current moves very slowly at low water (a few centimeters per second) upstream from this dam. Downstream, the depth of the river is reduced for a stretch of some 5 kilometers.

The Beautor Power Plant

This plant belongs to category (d) defined above. The headrace and tailrace canals and the section of river between them are respectively 1.07, 1.04, and 2.03 kilometers in length, resulting in a body of water with an over-all surface of approximately 120,000 square meters. Details:

Three 125,000-kilowatt sets (maximum intake rate: 13.5 cubic meters/sec)

Sited on the Oise, a tributary of the Seine

Rate of flow in severe low-water conditions: 5 cubic meters/sec

Main Difficulties Met in the Studies

A certain number of problems had to be resolved when these studies were undertaken. Some had been anticipated before the measurement process was started; others took us more or less by surprise. As much interest attaches to the description of these difficulties as to the statement of the results obtained.

Shortness of the Periods During Which Conditions Were Favorable for

Taking Readings

If the degree of cooling of the river is to be measured with adequate precision, the heating that has occurred must be considerable. In other words, the rate of flow must be low. The periods during which observations can be made last only one or two months and only occur once every three or four years (somewhat more frequently for some plants, less frequently for others). In practice, little useful measurement was possible in France in 1965, 1966, 1967, and 1968.

Instrumentation

It is useful to secure a continuous record of the temperature of the river at a certain number of points.

Robust recording thermometers capable of measuring the river temperature on a continuous basis exist, but they are not particularly sensitive (0.2° to 0.3° C.). These instruments operate by measuring the tension of a vaporized liquid.

Resistance thermometers are sensitive and give good results when used in factories, where they can be checked regularly; but disappointing results were obtained when they were installed and left unsupervised for periods of up to one week.

No fully satisfactory solution of this problem has yet been developed in France.

Influence of the Daily Temperature Cycle

In some experiments, an attempt was made to register the temperature at a certain number of points downstream from the plant, so as to cast light on the progressive rate of cooling of the river.

These measurements have to be taken over a stretch between 10 and 20 kilometers long and so must be spread over a large portion of the day. But at any given point, the temperature of the river will vary by about 0.5° C. as between night and day, independently of any effect from the water returned by the power plant. Thus, a limited degree of cooling only—and, indeed, sometimes, some rise in temperature—is noted as one moves slowly downstream between noon and 4:00 P.M. This phenomenon is a considerable drawback in interpreting the readings taken.

The difficulty, while a major one, could be eliminated either by setting up accurate recording apparatus at intervals of two or three kilometers along a section some 20 kilometers long (but we have not yet such fine instruments at our disposal), or by employing a large number of operators, each taking temperature readings at the same time at one or, at most, two locations.

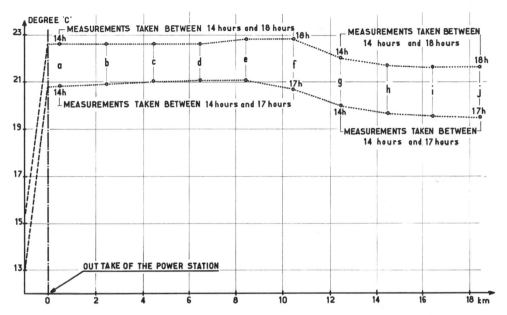

_BRIGHT SUNNY WEATHER
_VERY LIGHT WIND
_CONSTANT CAPACITY OF THE PLANT
_UNDER COVER AIR TEMPERATURE 23 °C
_TEMPERATURE OF THE RIVER 80 km DOWN STREAM 17.5 °C

Fig. 1——. Reading of temperature downstream from the Beautor Power Station, River Seine, France, on September 9 and September 21, 1966.

Determination of the Basic Temperature

It is often taken implicitly for granted that the temperature of the river upstream from the plant is its normal temperature and that the water warmed by the power plant will have resumed this temperature after perhaps a score of miles.

This is just not so.

The temperature of the Oise at Beautor, for example, is fairly low, and during the summer, the river warms naturally by 4° or 5° C. over the 80 kilometers downstream from Beautor. Thus, the water returned by the plant does not rejoin a river whose temperature is that observed upstream, but one whose temperature is higher and by an amount which is not known very accurately. The cooling process correspondingly appears to be slower than it really is, but the temperature observed upstream from the plant is never met again.

In exceptional cases, the temperature of the river after entry of the warm

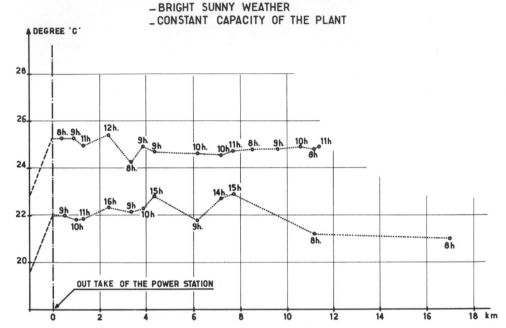

Fig. 2—. Reading of temperatures downstream from the Vaires-sur-Marne Power Station, River Seine, France, on June 24 and July 23, 1964.

flow from the power plant continues to rise downstream instead of falling.

Example 1 (see Figure 1): Water emerging from the power plant at 2:00 P.M. continues to heat further between the points (a) and (f), for the equilibrium temperature to which the river would normally tend during the afternoon is above the earlier downstream temperature, which itself is some 7° C. higher than the temperature upstream.

By contrast, the temperature difference between the points (a) and (g) or the points (f) and (j) reflects the decline in the river's downstream temperature from the plant occurring during the morning or the previous night (the current flowed at a rate of 20 kilometers per day during these series of readings).

Example 2 (see Figure 2): The water reheats slightly between 8:00 A.M. and 3:00 P.M. Depending on whether the equilibrium temperature is considered to be that upstream or is considered to have a rather higher value, the conclusion drawn will either be that the cooling process is only two-thirds complete after 20 kilometers, or that it is practically finished. This example shows how necessary it is to limit readings to periods during which heating occurs rapidly, i.e., during which the rate of flow is very low.

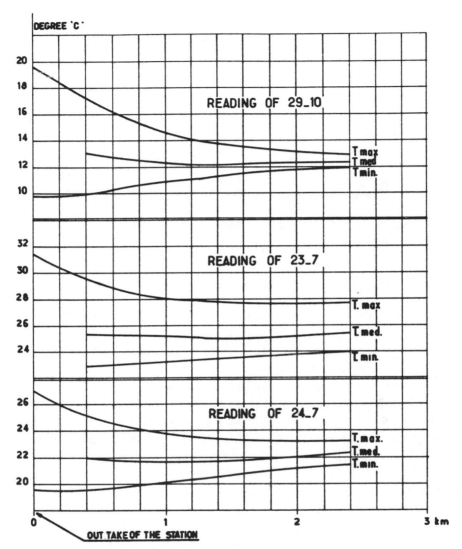

Fig. 3—. Results of three temperature readings made below the power plant at Voires-sur-Marne.

Determination of the Rate of Flow

Precise knowledge of the river's rate of flow at a different point is often indispensable in calculating the rate of cooling. This is an essential factor which should be analyzed with great care before taking any readings.

SECTION 1_ 24 JUNE 1964

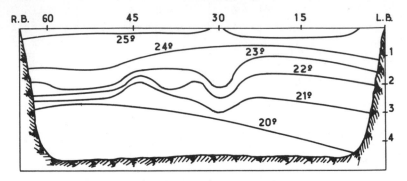

Fig. 4——. Thermal section of the River Marne.

RESULTS OBTAINED

Distance of Mix Downstream from the Plant

When warm water is returned to a slowly flowing and calm river, it merges with the cold water only with great difficulty; in particular, it tends to spread across the surface. This greatly hinders determination of the average temperature of the river.

Figures 3 and 4 illustrate this.

Figure 3 gives the results of three readings made below the power plant at Vaires. The origin of the abscissa is taken as the exit point of the used water from the plant. It takes almost 8 hours for the water to cover the 2,400-meter stretch over which readings were made. The average temperature appears to rise slightly, the farther one moves from the plant. This is the result of the phenomenon referred to above: to plot the curves, the temperatures were recorded over 4 different cross-sections and in each section at some 60 points. The readings were taken over a 6-hour period, moving downstream; correspondingly, there is a limited effect from the daily temperature cycle.

Figure 4 plots the temperature noted in a cross-section located 400 meters downstream from the plant (at this point, the river is some 60 meters wide and 5 meters deep). The marked horizontal stratification into layers of water of different temperatures is clear. It is of importance to note that boats do not navigate the Marne downstream from the plant, despite the depth of the river (they use a lateral canal). The case is similar, therefore, to that of a lake, which explains the clearness of the stratification and the length of the stretch over which the mingling of warm and cold water occurs.

In the case of the Seine downstream from Montereau, similar results are observed, except that the mix becomes homogeneous slightly more rapidly (perhaps because of the movement of vessels). As the Seine's rate of flow declines, the distance taken to complete the mix seems first to diminish, then to rise again (see table below).

TABLE 1. Temperatures and Flow-Rate, River Seine, near Montereau.

Rate of flow of Seine (cubic meter/sec)	250	115	65	55	38	28
Flow offtake by the power plant (cubic meter/sec)	10	10	10	10	10	10
Temperature difference at exit from plant (° C.)	6.9	7.0	6.2	6.9	6.8	7.3
Temperature difference 1400 meters downstream (° C.)	0.9	0.5	0.3	0.0	0.5	1.0
Time taken for water to travel this distance (hours)	0.7	1.5	2.5	2.9	4.2	5.7

Heat Dissipated by a Tract of Water

An attempt made to measure the rate of heat loss by the tract of water at the Beautor power plant was not very successful. The quantity of heat returned by the plant is known, and an estimate of the quantity of heat delivered downstream can be made by multiplying the rate of flow of the river by the rise in temperature between the intake point upstream and the exit point downstream from the plant. The difference between these two figures, where recirculation occurs, represents the heat evacuated by the tract of water formed by the headrace and tailrace canals and the section of river which separates them.

Over a set of five tests, the heat evacuated by the river downstream from the dam was estimated to represent 90, 91, 93, 94, and 97 percent of the heat output of the power plant, suggesting that the tract of water under study dissipated an average of some 8 percent. Unfortunately, an estimating error of 1 percent in assessing the rate of flow induces an error of 12 percent in the figure derived as the estimate of the tract's heat dispersal, and the accuracy with which the flow was measured was insufficient for the results of these series of tests to be considered as usable.

Cooling in a Canal or River

Let T_1 and T_2 be the excess over the equilibrium temperature in two sections of a canal (or river) a distance of x apart, L the width of the canal, Q its rate of flow. Then:

$$T_2 = T_1 \, exp(-\, kLx/Q)$$

in which expression, the coefficient k may vary with meteorological conditions (wind, temperature, humidity, etc.)

It should be noted that Lx in the expression represents the area of the water surface used to evacuate the heat.

Before going on to give the values of k determined experimentally, attention is drawn to a result which seems of some importance. The average increase in temperature at exit from the plant is A/Q (where A is the quantity of heat vented per second by the plant). At a distance x downstream, the maximum-temperature increase occurs for a flow equal to $Q = kLx$. This maximum increase, whose value is $T_{1/e}$, correspondingly is not necessarily the figure registered for the flow at low water. For example, 20 kilometers downriver from a 500,000-kilowatt station, for a river 100 meters wide, and for $k = 1.3 \; x \; 10^{-5}$, the maximum increase in temperature is found for a flow of $Q = 26$ cubic meters/sec, which may be considerably in excess of the flow at low water.

There is little point in reproducing here all the temperature readings which were collected; it will suffice to give the (very approximative) results derived from them. Taking as equilibrium temperature the temperature upstream from the plant (and stress has been laid on the extent to which this is an uncertain assumption), it would appear that at Vaires-sur-Marne, $k = 1.33 \; x \; 10^{-5}$. It is recalled that this coefficient was calculated for a calm river.

If this value of k is applied to a certain number of specific cases, we have, *for the tailrace canal of the Beautor plant:*

$L = 35.5 \; meters \quad x = 1040 \; meters$

$Q = 13.5 \; cubic \; meters/sec \qquad T_1 = 12.5 \; degrees \; Centigrade$ with $k = 1.33 \; x \; 10^{-5} \quad T_2 - T_1 = 0.46 \; degree \; Centigrade.$

Six values were found, ranging from 0.12° C. to 0.37° C., with an average of 0.28° C., which difference is not abnormal when it is recalled that the equilibrium temperature at Beautor is often higher than the temperature upstream from the plant. If this difference is 5° C., which appears a reasonable estimate, T_1 would then be 7.5° instead of 12.5° C., and the heat discharge calculated using the value $k = 1.33 \; x \; 10^{-5}$ would be 0.27° C., a result which is in line with experience.

For the *recirculation circuit at Beautor* (i.e., the headrace and tailrace canals and the stretch of river separating them), it may be calculated, taking $k = 1.33 \times 10^{-5}$, that the circuit ought to dissipate 15 percent of the heat output of the plant for a rate of flow of 8 cubic meters/sec, the capacity of the plant being 375,000 kilowatts. While experience indicates that the true value may be only 8 percent, notice has already been taken of the uncertainties affecting that figure. In addition, if the equilibrium temperature exceeds the upstream temperature, one would expect to record, in practice, less apparent cooling than is indicated by the theoretical calculated value.

For the cooling of *the Oise downstream from Beautor,* the figures in degrees Centigrade calculated using $k = 1.33 \times 10^{-5}$, using measures recorded over a distance of 9 kilometers, are given below. They assume that the equilibrium temperature is that observed a considerable distance downstream from the plant (and not upstream).

TABLE 2. Measured and Calculated Temperatures, River Oise, near Beautor.

	As measured	Calculated
9.9.1966	0.90	1.10
15.9.1966	0.57	0.65
21.9.1966	1.05	0.62

The temperature readings are very difficult to interpret, and the cooling figures given are, at best, very approximatively determined.

The cooling of the Seine downstream from Montereau: here, the calculation indicates a drop in temperature of 0.33° C., whereas the very rough readings made do not indicate any noticeable fall in temperature.

Figures for the cooling of *the Marne below Vaires-sur-Marne* for three stretches of river downstream from the plant (averages of three series of readings in 1964) are as follows:

TABLE 3. Measured and Calculated Temperatures, River Marne, near Vaires-sur-Marne.

As measured	Calculated
0.60	0.50
0.30	0.77
0.60	0.55

The Recirculation of Warm Water in a River

The warm water vented by a power plant into a canalized river spreads over the surface, and it is observed that some of it is drawn into the head-race intake upstream, even though the rate of flow of the river may exceed considerably the rate at which the plant pumps back its used water.

This phenomenon, technically very different from the phenomenon of cooling and much more complex, is at least as important to study. It will have an essential role to play in estuary installations.

Several simplified mathematical formulations have been advanced, but they provide, at best, a very imperfect representation of the phenomenon. Experimental research is a delicate matter. Electricite de France is considering the use of reduced-scale models to study the problems, based on the temperatures recorded at certain plants (in particular, Vaires-sur-Marne).

CONCLUSIONS

The measurements made in France in recent years have provided a certain number of results:

The mingling of warm and cold water downstream from the plant is a slow process which takes several kilometers before a practically homogeneous mix is obtained. However, because of the quite regular stratification of the water into layers, it seems fair to assume that the high temperature at which water issues from the power station's condensers is not particularly troublesome to the fish in the river; so far as pisci-culture is concerned, the relevant parameter is the temperature obtaining after the mixing process is completed.

In my view, there is no longer any need in France to take readings for canalized rivers; I would merely draw attention to the fact that the temperature increase resulting from passage through the condensers is almost always less than $8°$ C.

Figures have been obtained for the cooling occurring downstream from the plant, but they are only orders of magnitude, which cannot as yet be used for making sufficiently accurate projections. But we have also seen the major difficulties attached to practical experimentation: the perfecting of instrumentation for taking temperature, the need to use large quantities of manpower during a short interval of time to measure both temperatures and rates of flow.

It would seem that progress may lie in either of two very different directions. One is to undertake a series of discontinuous measurements (having kept the power plant operating at as nearly constant a rate as possible during two or three days, make simultaneous readings of the temperature of 10 or 15 sections of about 2 kilometers each). The other is to take continuous readings (the temperatures in two sections are measured continuously, and an attempt is made to calculate the energy balance sheet for the body of water lying between them). The first approach carries certain obligations with it (the need to keep the power plant functioning at a constant rate, the important inputs of instrumentation and personnel), at the same time as such accidental occurrences as heavy rainstorms or plant failures may forfeit the benefit of any given measurement campaign. The second approach involves obtaining more precise knowledge of temperatures and rates of flow, but it does provide some hope of showing the influence of such parameters as wind speed or amount of sunshine. On the other hand, it can give no indication of the effect of the specific geographical configuration of the section for which measurement is envisaged (river passing through a forest or urban area, riverbed frame, etc.).

Doubtless the only workable solution will lie in some combination of the two approaches.

DISCUSSION   John Eric Edinger

THE heat disposal problem can be divided into two parts which are analyzed almost separately. First is the initial mixing or dilution of the discharge; second is atmospheric cooling. These two parts have different length- and time-scales. Mixing takes place in the immediate vicinity of the discharge and affects a small portion of a waterbody. Atmospheric cooling is relied on after mixing and is a long-term process.

Dilution of power plant discharges, like other dilution problems, implies that there is water available for mixing. With a 10° F. or 15° F. temperature rise across a condenser, dilution water must be available up to ten to fifteen times the volume of discharge. These are small dilution ratios when compared to the 100- to 500-fold requirements of sewage discharges. The large condenser flow rates discharged at fairly reasonable velocities have a very high momentum and are capable of entraining large quantities of cold water, if that cold water is available.

As an example, measurements have shown that a surface discharge off the Pacific Coast from a 1000-mw plant entrains sufficient cold water to decrease the initial temperature rise of 10° F. to less than 1° F. within a surface area of about 200 acres (Cheney and Richards, 1965). In this case, large volumes of dilution water were available and were drawn into the 200-acre discharge area by entrainment. For comparison, if the same discharge underwent no mixing, it would have required approximately 2300 acres of surface area to achieve the same reduction in temperature by surface cooling alone.

On a river, the water available for dilution is that flowing past the plant discharge. There is reason to believe that momentum principles can be applied to analyze the rate of mixing of the discharge up to full dilution with the river flow. This problem of initial mixing of discharges in rivers has been almost ignored, up to the present time. It is a problem that will

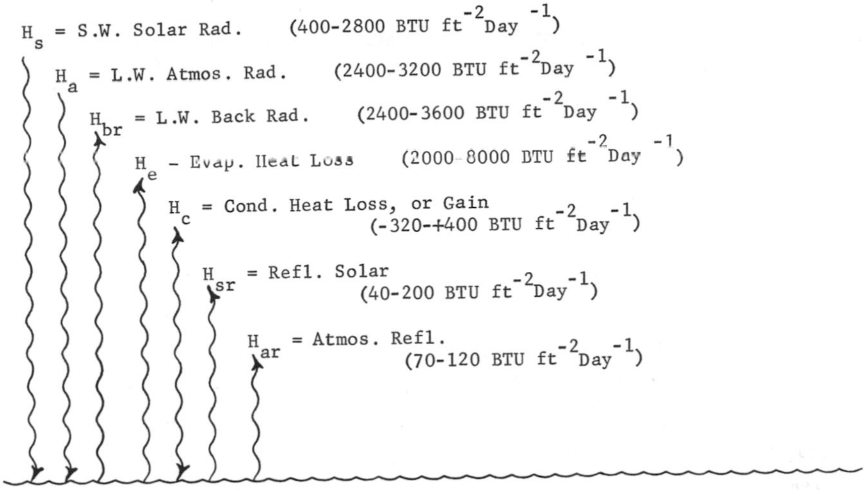

H_s = S.W. Solar Rad.   (400-2800 BTU ft^{-2}Day^{-1})

H_a = L.W. Atmos. Rad.   (2400-3200 BTU ft^{-2}Day^{-1})

H_{br} = L.W. Back Rad.   (2400-3600 BTU ft^{-2}Day^{-1})

H_e – Evap. Heat Loss   (2000-8000 BTU ft^{-2}Day^{-1})

H_c = Cond. Heat Loss, or Gain
                 (-320-+400 BTU ft^{-2}Day^{-1})

H_{sr} = Refl. Solar
                 (40-200 BTU ft^{-2}Day^{-1})

H_{ar} = Atmos. Refl.
                 (70-120 BTU ft^{-2}Day^{-1})

NET RATE AT WHICH HEAT CROSSES WATER SURFACE

$$H_n = \underbrace{(H_s + H_a - H_{sr} - H_{ar})}_{\substack{\text{Absorbed Radiation,}\\ \text{Independent of Temp., } H_R}} - \underbrace{(H_{br} \pm H_c + H_e)}_{\text{Temp. Dependent Terms}} \text{ BTU ft}^{-2}\text{Day}^{-1}$$

$$H_{br} \sim (T_s + 460)^4$$

$$H_c \sim (T_s - T_a)$$

$$H_e \sim W(e_s - e_a)$$

Fig. 1.—Mechanisms of heat transfer across a water surface.

have to be studied if any ability is going to be developed in prediction of temperatures immediately around power-plant outfalls.

After initial mixing, the usual advection, dispersion, and cooling processes control temperature distributions. It is in the study of these that most recent progress has been made.

A common problem encountered in the study of condenser-cooling water discharges is determination of the temperature distribution that would result from increasing the size of an existing steam plant. Analyzing the existing distribution of temperatures that surround a power plant is an initial step toward temperature predictions. In the analysis of cooling-water discharges, continuity relationships are applied to the temporal and spatial distribution of heat with exchange at the water surface superimposing an additional term. When surface-heat exchange is introduced into the continuity relationship, it becomes analogous to reaeration in the oxygen-continuity relations, or chemical reaction and radioactive decay in the continuity equations for mass.

The rate of heat exchange at the water surface can be formulated from the term-by-term heat budget as shown in Figure 1. The net rate of heat exchange is defined as the algebraic sum of the rates at which heat is transported across the water surface by short-wave solar radiation; long-wave atmospheric radiation; reflected short- and long-wave radiation; back radiation from the water surface; evaporation; and conduction (Edinger and Geyer, 1965). The surface-heat transfer terms can be divided into two groups: the absorbed-radiation terms, which are independent of water-surface temperature; and the terms which are dependent on water-surface temperature. When the back radiation, evaporation, and heat conduction are expressed in terms of their physical dependence on water temperature, air temperature, and air-vapor pressure, the net rate of heat exchange becomes proportional to the difference in water-surface temperature and equilibrium temperature, as:

$$H_n = -K(T_s - E) \qquad \text{(Equation 1)}$$

where H_n is the net rate of surface-heat exchange, K is the thermal-exchange coefficient, T_s is the water-surface temperature and, E is the equilibrium temperature. Further, the equilibrium temperature is formulated from the above considerations to be, in approximate form:

$$E = \frac{H_R - \varepsilon\sigma\Delta^4}{K} + \frac{K - 4\varepsilon\sigma\Delta^3}{K(C_1 + \beta)} (C_1 T_a + \beta T_d) \qquad \text{(Equation 2)}$$

where, H_R is the absorbed radiation (Figure 1); ε is the emissitivity of water; σ is the Stephan-Boltzmann constant; Δ is the conversion to absolute temperature; C_1 is the Bowen constant in the Bowen ratio; β is the slope of the temperature-saturation vapor-pressure relation; T_a is air temperature, and T_d is the atmospheric dew-point temperature.

Thus, the equilibrium temperature is the water temperature at which there is no heat exchange across the water surface and it is a function of short-wave and long-wave radiation, air temperature, and air-vapor pressure. The equilibrium temperature is continually changing. The actual water

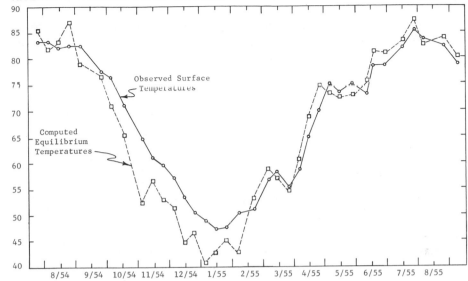

Fig. 2——. Equilibrium and surface temperatures and exchange coefficients for Lake Colorado City, Texas.

temperature at any instant is being driven toward the equilibrium temperature at a rate proportional to difference between them.

It should be emphasized that the equilibrium temperature is derived from the term-by-term heat budget and is a function only of meteorological variables. Basically, it has three components; one due to solar and atmospheric radiation; one due to air temperature; and one due to air-vapor pressure. The equilibrium temperature thus contains as much information as the term-by-term heat budget shown previously. It is, however, easier to insert into temperature-prediction equations and is a particularly useful concept when studying temporal variations of water temperatures.

A typical response of natural water temperatures to the equilibrium temperature is shown in Figure 2. These figures are for a seasonal cycle of ten-day averages and show that, as heat is being lost, or as the lake is cooling down, the water temperatures are above the equilibrium temperature. The reverse is seen when the lake is warming up. Note, in this case of seasonal temperature variations, that the amplitude of the actual water temperatures and equilibrium temperature are similar in magnitude. The diurnal case will be presented shortly.

Water temperatures in a waterbody used as a sink for waste-heat discharges exhibit diurnal fluctuations in response to the heat load from the plant and in response to varying meteoroligical conditions. The amplitude of the diurnal water-temperature variation is largest near the plant and decreases to approach the amplitude of the naturally occurring water temperatures far away from the point of discharge. The time of occurrence of peak daily temperatures varies throughout the waterbody, depending on the distance or travel time from the plant, as well as on the time of day.

As indicated by Goubet (1968), an examination of the daily water-temperature cycle is most important for a flowing stream in which steam plants are located in series. In such situations, the time and magnitude of temperature effects at each downstream intake must be known. It is also important to be able to separate the temperature variations due to changes in power production from the variation due to changes in solar radiation and other meteorological conditions. Development of relationships which predict the temperature distributions due to changes in capacity and operation is obviously desirable.

The application and use of these concepts is best illustrated by a case study (Edinger, Brady, and Graves, 1968). A power-plant discharge situation selected for discussion of the influence of heated discharges on the temporal variation of water temperatures is shown in Figure 3. The lake was constructed specifically for the purpose of dissipating the waste heat before returning the cooling water to a nearby stream. It is a simple flow-through cooling pond, in which water temperatures are continuously recorded at a number of points along its axis.

The series of water-temperature data chosen for analysis is shown in Figure 5. The daily cycle of water temperatures is uniform over a six-day period, with no increasing or decreasing trend in average daily temperatures. This is the simplest type of record for which to show the pertinent characteristics of a time-varying discharge. There is a general decrease in the mean weekly temperature proceeding from Station B, near the point at which the plant flow enters the lake, to Station H, at which the plant flow leaves the lake. This represents the dissipation, or loss, of heat from the flow-through pond.

The amplitudes decrease from a weekly average of 12.5° F. at Station B to approximately 2° F. at Station H. The time of occurrence of peak temperatures at Station B is near 1200 hrs, which corresponds to the time of maximum plant loading; and at Station C, the peak temperatures occur near 1800 hrs, indicating a time-lag of flow from Station B to Station C. It is difficult to associate the peak temperatures at Stations G and H with those at the plant for an earlier time, because the response to the meteorological conditions becomes more significant.

The weekly mean temperature, average of the daily maximum tempera-

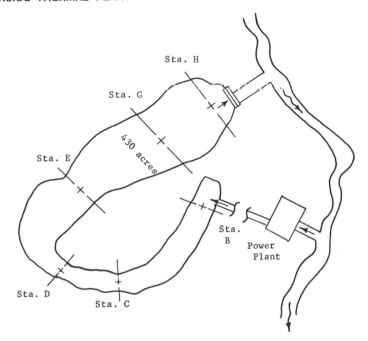

Fig. 3——. Schematic of Site No. 3 Cooling Lake.
SOURCE: From Edinger, Brady, and Graves, 1968.

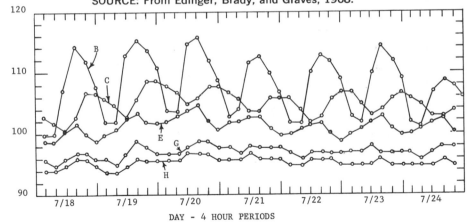

Fig. 4——. Observed temperatures, Site No. 3, July 18–July 24, 1966.

tures computed from the time-series data at each station, are shown as
profiles of temperatures along the axis of the pond in Figure 5. These observations indicate that the chosen characteristics of the water-temperature
records decay along the pond in a manner described by solutions to the
one-dimensional time-varying equation for continuity of heat. Fitting of

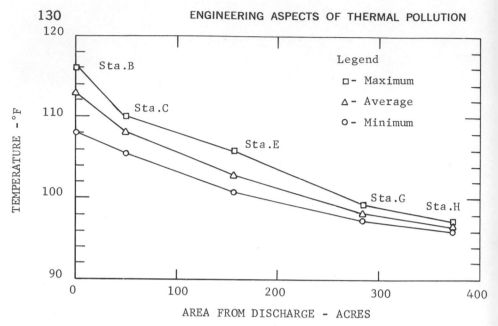

Fig. 5——. Longitudinal profile of averaged maximum, mean, and minimum temperatures, Site No. 3, July 18–July 24, 1966.

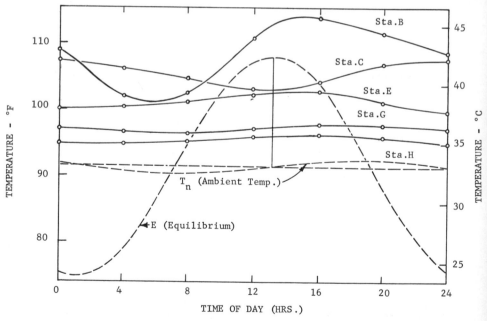

Fig. 6——. Comparison of computed equilibrium and ambient temperatures with observed mean diurnal temperature variations for Site No. 3, July 18–July 24, 1966.

the observations to the equations for the weekly period yields a thermal-exchange coefficient of 174 BTU $Ft^{-2}Day^{-1}{}^\circ F^{-1}$, and a mean weekly amplitude for the ambient water-temperature variation of approximately 3° F.

The equilibrium-temperature mean value and amplitude can be determined directly from the temperature records, independent of the meteorological data, by assuming it has a diurnal sinusoidal variation. Its variations can be compared with the meteorological data in future analyses. The sinusoidal equilibrium-temperature variation is shown in Figure 6, superimposed on a plot of the mean diurnal temperature variations at Stations B, C, E, G, and H. It has a mean value of 91.5° F. and an amplitude of 33.6° F. The time-lag of the maximum ambient temperature from the maximum equilibrium temperature is 5.5 hours. Comparing this diurnal picture with the seasonal picture shown earlier, it is seen that the relative amplitudes of the equilibrium temperature and the actual water temperatures is larger for the diurnal cycle than for the seasonal cycle. Also, the lag-time between the peak equilibrium temperature and the ambient water temperature is relatively longer for the diurnal amplitude of the equilibrium temperature are similar in magnitude. These properties are primarily due to the influence of short-wave solar radiation.

Returning to the diurnal problem illustrated in Figure 6, it is seen that the equilibrium temperature reaches its peak some time after 12 noon, due to the difference between solar time and local time of day. The maximum equilibrium temperature has a value of 108.3° F., which is exceeded by the peak discharge temperatures recorded at Station B. For the period of observation there is a net loss of heat throughout the whole day for most of the surface area between Station B and Station C. Heat is being lost from the water surface between Station C and the end of the pond during the evening and early morning hours, but there is a net heat gain from the atmosphere and solar radiation during late morning and afternoon. The maximum temperatures at Stations G and H tend to occur late in the afternoon, but the mean values of these temperatures remain above that of the computed ambient temperature.

REFERENCES

Cheney, W. O., and G. V. Richards. 1965. "Ocean Temperature Measurements for Power Plant Design." *ASCE Santa Barbara Conference on Coastal Engineering,* October 1965.

Edinger, J. E., D. K. Brady, and W. L. Graves. 1968. "The Variation of Water Temperatures Due to Steam-Electric Cooling Operations." *J. Water Pollution Control Federation* 40(9):1637–1639.

Edinger, J. E., and J. C. Geyer. 1965. *Heat Exchange in the Environment*. New York: Edison Electric Institute Publication 65–902.

Goubet, A. 1968. "Prediction of Heat Dissipation in Rivers, Reservoirs, and Lakes." Paper read at National Symposium on Thermal Pollution: Engineering and Economic Considerations, August 14–16, 1968, at Nashville, Tennessee. (See pp. 110–123, this volume.)

DISCUSSION   G. Earl Harbeck, Jr.

EXPERIENCE IN FRANCE

MR. GOUBET has provided a lucid and interesting description of the cooling-water problems associated with three French thermal-electric plants. Most of the time, as the author mentions, flow rates are sufficiently high that temperature rises are small and difficult to measure. Only under infrequent conditions of drought-flow is it possible to obtain well-defined temperature profiles below the plants in order to attempt to delineate the temperature die-away patterns. In addition, his comments on the accuracy of continuous-temperature recorders are well taken. Only by frequent and careful comparisons with a good standard thermometer can reasonably accurate data be obtained. Moreover, it is difficult to be sure that temperatures recorded at a single point are representative of the entire stream.

Mr. Goubet noted a diurnal range of water temperature of about $0.5°$ C. Data to be presented by the writer for two U.S. streams show a diurnal variation of between $1°$ and $2°$ C., but presumably the normal flow of these streams is not nearly as large as the flow of those studied by Mr. Goubet.

He also noted that warm water, when added to a slowly flowing, calm river, tends to spread across the surface, thus hindering determination of the average temperature of the river. This is true but is actually an advantage. The heat added by the plant is returned to the atmosphere through surface processes; that is, the higher the surface temperature, the more heat dissipated. Theoretically, it would be most desirable to add the warm water as a thin sheet, avoiding all mixing if possible. This would result in the most rapid dissipation of heat to the atmosphere.

Mr. Goubet presented an equation to describe the downstream decrease in temperature, as follows:

$$T_2 = T_1 \, e^{-kLx/Q}$$

133

in which T_1 and T_2 are the excess over the equilibrium temperature in two sections of a canal (or river) a distance of x apart, L is the width of the canal, and Q is its rate of flow. The coefficient k may vary with meteorological conditions. The author is correct in his intuitive choice of the mathematical form of equation he has selected. It can be reasonably expected that the downstream temperature pattern will be similar to that defined by an exponential die-away. But the return to the atmosphere of heat added by the plant is accomplished through three important but separate processes, and although their combined effect does approximate an exponential die-away, as described later in this paper, there is no obvious reason, at least to the writer, why this should be so. More research is needed.

The author concludes that there is no longer any need in France to take readings for canalized rivers, in view of the fact that the temperature rise through the plant is almost always less than 8° C. Perhaps this might be true also of most United States plants, but there are exceptions. Data will be presented later for the West Branch of Susquehanna River, where the maximum temperature rise during a 24-hour study period in 1962 was 15° C. I think most engineers and hydrologists will agree that a much smaller temperature rise would be desirable, but is not always attainable.

Mr. Goubet's paper provides some interesting data on the thermal-pollution problem in France, and suggests two plans of study. The first would involve discontinuous measurements at some 10 or 15 downstream sections over a period of two or three days, during which the plant remains on constant load. The second requires continuous records at both ends of a single reach. Probably he is correct in his statement that the only workable solution will lie in some combination of the two approaches.

EXPERIENCE IN THE UNITED STATES

Theory

One of the first to advance cooling-pond theory probably was Throne (1951), who had a good understanding of the basic principles involved but unfortunately lacked the equipment necessary for testing the theory completely. The most important deficiency was the lack of instrumentation for measuring incoming atmospheric, or long-wave radiation. This deficiency was remedied by the invention of the total hemispherical radiometer (Dunkle and others, 1949); and soon thereafter, the energy-budget method of measuring reservoir evaporation was tested by the United States Geological Survey (1954) and other federal agencies at Lake Hefner in 1950–51.

At the request of the United States Bureau of Mines, the writer studied the problem of disposing of unwanted heat and published a short report (Harbeck, 1953) which went a step further than Throne's earlier basic work, in that it utilized both energy-budget and mass-transfer theory to permit determining two unknowns, namely, the increase in evaporation and the rise in water-surface temperature resulting from the addition of heat by a power plant or other source. The method was tested at Lake Colorado City in 1954–55 (Harbeck, Koberg, and Hughes, 1959) and again in 1959–60 (Harbeck, Meyers, and Hughes, 1966).

The relatively simple theory has been utilized independently by a number of other investigators, including McLean.(1961) and Raphael (1962). Disregarding certain minor items, the energy budget for a reservoir or stream can be written as follows:

$$R - B - E - H - G + C = S \qquad \text{(Equation 1)}$$

in which $R =$ net incoming radiation, both long- and short-wave
$B =$ long-wave radiation emitted by the water surface
$E =$ energy utilized for evaporation
$H =$ energy conducted to the atmosphere as sensible heat
$G =$ energy conducted into the reservoir bottom stream bed
$C =$ net energy brought in as inflow or taken out as outflow, including heat added by the plant
$S =$ increase in energy stored in the body

In Equation 1, H must be computed using the Bowen ratio concept because equipment for its direct measurement under field conditions is not routinely available. Bowen assumed that the diffusivities of heat and water vapor were equal, which is highly questionable under the tremendously unstable lapse rates existing where heat loads are large, causing air-water temperature differences to be great. Among the items in Equation 1, it is obvious that R is not affected by the addition of heat by a plant. If we assume that, after the initial increase in S occurs because of the plant, the seasonal curve of change in energy storage parallels the natural seasonal change, then the effect of the addition of heat by a plant has a negligible effect upon S. This latter assumption also implies that the plant load remains relatively constant. In a combined steam-hydro system, however,

modern practice is to keep the steam plants on relatively constant load, re-
lying on hydro for peaking power, so that the assumption is not entirely
unreasonable.

Thus for a reservoir to which heat is being added,

$$\triangle B \; + \; \triangle E \; + \; \triangle H \; + \; \triangle G \; = \; \triangle C \qquad \text{(Equation 2)}$$

in which $\triangle$ represents the change in each item resulting from the addition
of heat by the plant.

The mass-transfer equation is

$$E = LNu(e_o - e_a) \qquad \text{(Equation 3)}$$

in which $E =$ evaporation

$L =$ latent heat of vaporization

$N =$ a coefficient of proportionality, often called the "mass-
transfer coefficient"

$u =$ wind speed

$e_o =$ saturation vapor pressure corresponding to the temperature
of the water surface

$e_a =$ vapor pressure of the ambient air

The mathematical and physical relationships involved in Equations 2
and 3 are well known, and need not be repeated here. The use of these
two equations, involving both energy-budget and mass-transfer concepts,
permits the determination of the two unknowns, namely, the amount of
forced evaporation and the water-temperature rise resulting from the addition of heat by the power plant. For reservoirs, particularly in arid regions,
it is desirable, if not essential, that both of the previously mentioned unknowns be determined. In humid regions, the increase in evaporation may
be of only academic interest, but the magnitude of the temperature increase appears to be of concern everywhere.

Although the theory was derived for use in studies of reservoirs or lakes,
there is no reason why it cannot be used for streams as well. Consider a
reach of river below a source of heat. Divide it, if you will, into a half-
dozen subreaches. Each can be thought of as a small reservoir, with inflow
at the upstream end and outflow at the downstream end. A balanced energy
budget using Equation 2 can be computed for each reach. Using the observed temperature at the upper end of the first subreach as a point of
beginning, it is possible to compute a temperature at the lower end of that

subreach that will cause Equation 2 to balance. The computed temperature at the lower end of the first subreach becomes, then, the temperature at the upper end of the second subreach, and the process is repeated until the temperature die-away pattern is described adequately.

The Holston River Study

The Holston River field study was made in November 1961, as a co-operative effort of the Tennessee Valley Authority and the Geological Survey. The site was a reach of Holston River below the John Sevier Steam Plant near Rogersville, Tennessee. Most of the year, this reach of the river is part of Cherokee Reservoir, but in the fall, the reservoir is usually drawn down in anticipation of winter flooding, and at this time, flow through the reach is turbulent.

TVA maintained a constant plant load during a period of more than 24 hours, during which the entire flow of the river, about 640 cfs, was taken through the plant. This is not the normal operating procedure; ordinarily, only part of the river-flow is put through the plant. For the test period, however, most of the normal river-flow was stored in an upstream reservoir, so that a much greater-than-normal temperature rise could be obtained.

A complete meteorological station for determining evaporation by the energy-budget method was installed at Rogersville, about three miles from the nearest point of the river reach. The equipment included a pyrheliometer, a total hemispherical radiometer, and wet- and dry-bulb thermocouples. The various outputs were recorded on a multi-channel strip-chart potentiometer. In addition, the outputs were totalized each hour, using a mechanical integrating system devised by C. R. Daum of the Geological Survey. This last luxury eliminated the need for laborious processing of the recorder chart; it was necessary only to read the integrator dials hourly. The chart was scanned, of course, to detect any possible instrumental malfunctions—none occurred. Wind speeds were measured at a height of between one and two meters above the water surface at three places that were selected as having representative exposures.

Flow of Holston River was measured, using conventional stream-gaging techniques, at the point of discharge from the plant and at the lower end of the 11-mile reach. Tributary inflow in the reach was measured but was negligibly small.

The plant inflow- and outflow-temperatures were measured, and important water-temperature data were obtained at six sections. The first was 0.6

miles below the plant, and the others were 2 to 2½ miles apart. At each section, a vertical profile of water temperature was obtained at each of 10 points across the river. Continuous records of water temperature at a selected point were obtained at three of the six sections.

One item in the energy budget that was not measured was the flow of heat from the river into its underlying sediments. The writer takes the entire responsibility for this omission. At first glance, it did not appear to be significantly large, because, presumably, equilibrium had been reached and temperature gradients small. Such was not the case. Equilibrium had not been reached, simply because conditions during the test were not representative. Ordinarily, only part of the river-flow goes through the plant, and the rise in temperature of the entire river is much less than during the test. Thus, when the test began, river-flow decreased to about 640 cfs, and its temperature increased markedly, so that there must have been a substantial flow of heat into the bottom sediments. These sediments are of appreciable thickness and are soft, as this reach of river is normally a part of the upstream end of the reservoir, and much of the sediment load of the river is deposited here. An estimate was made of the heat flow into the bottom sediments, based on Ingersoll's heat-flow theory and some estimated thermal conductivities and diffusivities, but the results are not considered particularly accurate.

The Susquehanna River Study

A second study was made by the Geological Survey, using a reach of the West Branch of the Susquehanna River below Shawville, Pa., and a report on the study was prepared by Messinger (1963). The power plant, which operates at a constant load of 600,000 kw, raised the temperature of the river about 15° C. during a 24-hour period in October 1962. Flow of the river was virtually constant at 370 cfs.

The field procedures were essentially the same as those used on the Holston River and need not be described again. In addition, temperatures of stream-bed materials were measured, using a hollow plastic probe containing three thermistors embedded six inches apart which were connected to a portable Wheatstone bridge. In contrast to Holston River, however, the West Branch of the Susquehanna has a rocky bed with only a thin, partial sediment cover. The thermal conductivity of the bottom material was not measured. Handbook values were used, but the possible error from this source was small, since the heat loss to the stream bed was only about 5 percent of the sensible heat loss in each subreach.

In his analysis, Messinger assumed that the ratio of the diffusivity of heat to that of water vapor was about 1.3, as compared with Bowen's (1926) assumption of equality. It seems to be agreed that this ratio should be greater than unity when the atmospheric-lapse rate is unstable (as it usually is over heated water), but there is no general agreement as to how much greater it should be. Messinger's assumption was unquestionably theoretically sound, in a qualitative sense, but it remains to be seen whether it is also quantitatively correct.

Results of the Two Studies

The variation of stream temperature with time at each of the downstream sections is of considerable interest. The superpositions of the natural diurnal temperature pattern upon the plant-caused pattern is clearly evident.

On the Holston River (Figure 1), the reason for the decrease in plant temperature at about noon on November 27 is not known. Note that this sag is barely evident at Section 1, 0.6 mile downstream. Both the plant intake and discharge temperature reached a maximum at about 1600 on the twenty-seventh; presumably the lateness of the diurnal peak was related in some way to the effect of the small storage pond above the plant.

Although the plant went on constant load at 2100 on November 26, it is apparent that equilibrium was not established at Sections 3 and 4 until sometime after observations began at 0800 on the twenty-seventh.

On the West Branch of the Susquehanna River, the diurnal effect was considerably greater (Figure 2). It is clearly evident in the record of inflow to the plant and at all the downstream sections. Because the Shawville plant normally operates at constant load, conditions in the river were at relative equilibrium.

The results of the two studies are shown in Figure 3. Note that the shapes of all curves approximate the exponential die-away postulated by Mr. Goubet. The major portion of heat added to a stream is disposed of through three separate processes, and it is not clear why their sum should approximate an exponential, but obviously it does. Note also that the rate of downstream decrease in temperature computed as described earlier in this paper is considerably less than was observed. Messinger (1963, p. C–176, 177) attributed this to three causes, in decreasing order of probability, as follows:

1. Measured solar radiation may be considerably greater than that actually reaching the stream because of shading by nearby hills and trees.

2. Errors in measured wind speeds, which were often less than the threshold response of the anemometers.

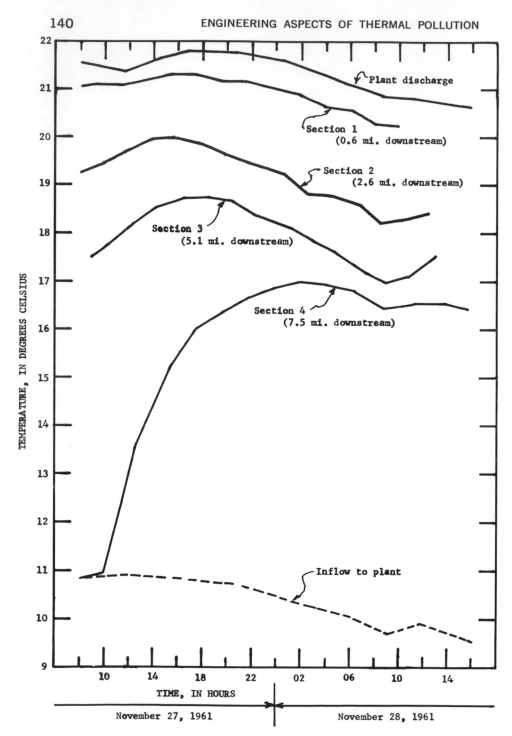

Fig. 1—. Water temperature below John Sevier Power Plant on Holston River, Tennessee.

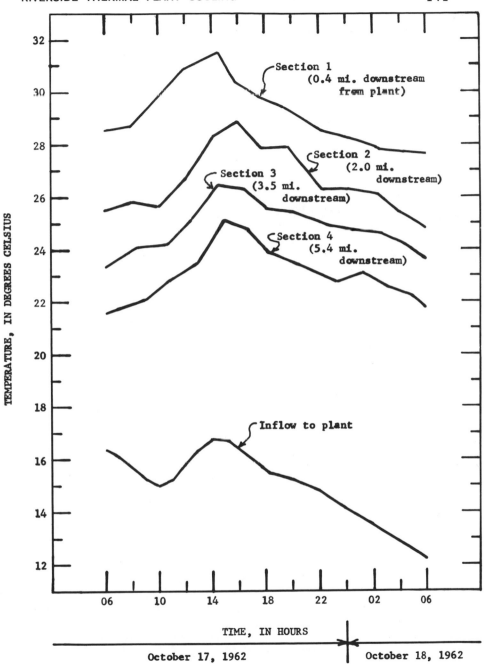

Fig. 2—. Water temperature below Shawville Power Plant on west branch of the
Susquehanna River, Pennsylvania.
SOURCE: After Messinger, 1963.

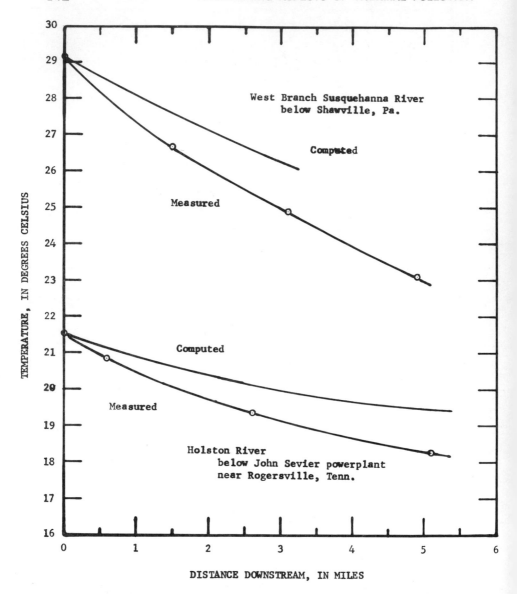

Fig. 3——. Comparison between computed and measured downstream temperatures on Holston River and west branch of the Susquehanna River.

SOURCE: Susquehanna River data after Messinger, 1963.

    3. Water-surface temperatures may have been measured inadequately, and this item is of great importance in the energy-budget calculations.

Some light may be shed on the third point by the use of airborne remote-sensing equipment. With adequate "ground truth," infrared imagery may show whether downstream temperature die-away patterns are defined sufficiently well by the use of selected cross-sections.

Another point that has been mentioned previously is the uncertainty concerning the equivalence of the diffusivities of heat and water vapor when the lapse rate is unstable. A comparison between the measured and computed temperature die-away patterns shows that much more heat is being returned to the atmosphere than theory indicates, and it is quite possible that some, if not most of it, is being returned by forced convection. In any event, research is needed.

REFERENCES

Bowen, I. S. 1926. "The Ratio of Heat Losses by Conduction and by Evaporation from any Water Surface." *Physical Review* 27:779–787.

Dunkle, R. V., and others. 1949. "Nonselective Radiometers for Hemispherical Irradiation and Net Radiation Interchange Measurements." *Thermal Radiation Project Report 9,* California University Engineering Department.

Harbeck, G. E., Jr. 1953. "The Use of Reservoirs and Lakes for the Dissipation of Heat." *U. S. Geological Survey Circular 282.* [Washington, D.C.: U.S. Gov't. Printing Office.]

Harbeck, G. E., Jr., G. E. Koberg, and G. H. Hughes. 1959. "The Effect of the Addition of Heat from a Power Plant on the Thermal Structure and Evaporation of Lake Colorado City, Texas." *U.S. Geological Survey Professional Paper 272–B.* [Washington, D.C.: U.S. Gov't. Printing Office.]

Harbeck, G. E., Jr., J. S. Meyers, and G. H. Hughes. 1966. "Effect of an Increased Heat Load on the Thermal Structure and Evaporation of Lake Colorado City, Texas." *Texas Water Development Board Report 24.*

McLean, J. E. 1961. "Thermal Pollution of Streams." Unpublished term paper, Harvard University.

Messinger, H. 1963. "Dissipation of Heat from a Thermally Loaded Stream." *U.S. Geological Survey Professional Paper 475–C,* Article 104, C175–C178. [Washington, D.C.: U.S. Gov't. Printing Office.]

Raphael, J. M. 1962. "Prediction of Temperature in Rivers and Reservoirs." *Journal of the Power Division, Proceedings of the American Society of Civil Engineers,* 88 (July): PO2.

Throne, R. F. 1951. "How to Predict Lake Cooling Action." *Power 95* (Sept.): 86–89.

U. S. Geological Survey. 1954. "Water-Loss Investigations—Lake Hefner Studies: Technical Report." *United States Geological Survey Professional Paper 269.* [Washington, D.C.: U.S. Gov't. Printing Office.]

Chapter **5**   Donald R. F. Harleman

MECHANICS OF CONDENSER-WATER DISCHARGE FROM THERMAL-POWER PLANTS

THIS paper is concerned with the disposal of heated condenser water from thermal or nuclear-power plants into an adjacent waterway. It is not proposed to argue the relative merits of heat dissipation in waterways versus dissipation directly to the atmosphere. It is recognized that, when alternative solutions are available, it is good engineering practice to analyze and make predictions of environmental changes which are to be expected. The ultimate choice depends upon the relative benefits and costs of the various alternatives. In considering whether all or a portion of the total waste-heat is to be disposed of into an adjacent waterway, use should be made of the best theoretical and experimental techniques in the planning and design stage. The objective of this paper is to discuss some of these techniques and to show the degree of flexibility in controlling conditions in the waterway which is available to the engineer. This flexibility is best illustrated by considering the two ends of the spectrum of heat disposal into an adjacent waterway: (a) stratification of the condenser-water discharge and (b) complete mixing.

Condenser-water discharge structures can be designed to produce almost complete stratification or complete mixing. Intermediate conditions may also be achieved, although quantitative design methods are not, as yet, well developed.

STRATIFICATION OF CONDENSER-WATER DISCHARGE

To achieve stratification of a heated condenser-water effluent in an adjacent waterway, the condenser-water outlet structure must be designed

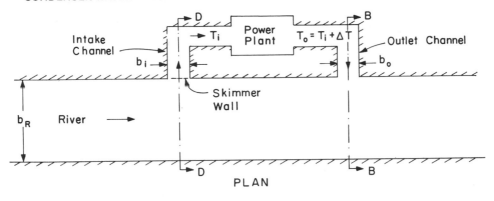

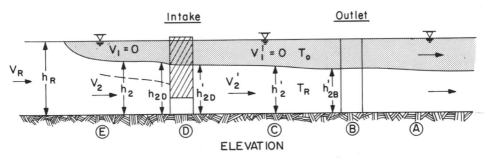

Fig. 1—. Flow stratification in vicinity of power plant.

to minimize mixing between the heated effluent and the receiving water. In general, this results in a heated surface layer, both upstream and downstream of the outlet channel. Heat dissipation to the atmosphere is at the highest rate, since the surface-layer temperature is a maximum. Recirculation of the condenser water at the upstream intake can be controlled by means of a skimmer wall.

A schematic diagram of a typical power-plant site on the bank of a river is shown in the plan view of Figure 1. Water withdrawn from the river enters the plant at a temperature T_i and leaves the plant with a temperature $T_o = T_i + \triangle T$. If surface-heat losses in the intake and outlet channels are neglected, and if recirculation is effectively prevented, $T_i = T_R$, where T_R is the ambient river temperature. The stratification of the heated condenser-water discharge in the river adjacent to the power plant is shown in the elevation section of Figure 1. Five zones of flow are designated by the letters A through E proceeding in the upstream direction and starting with the portion of the river immediately below the outlet channel.

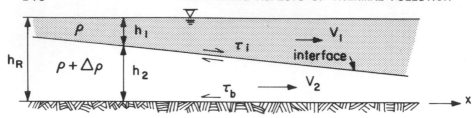

Fig. 2——. Two-layered stratified flow.

In some instances, water-temperature regulations may prevent use of this method, since the temperature of the surface layer may be above permissible values in the summer. However, the effect on the ecology of the waterway may not be severe. The heated water is confined to a relatively thin surface layer, and the region of temperature increase tends to be localized by the high rate of heat loss from the water surface. In addition, condensers of large volumetric capacity may be employed to reduce the temperature rise through the power plant.

The analysis of the non-mixing effluent is based upon the theory of stratified flow, a branch of fluid mechanics which deals with the effect of small density differences on fluid motion in layers. A density difference between layers of the order of a tenth of one percent, corresponding to a temperature difference of 6° F., is large enough to be of significance in a stratified flow.

Stratified-Flow Theory

A two-layer stratified-flow system is shown in Figure 2. The flows are assumed to be steady, and each layer is homogeneous, with a density differential $\Delta\rho$ across the interface between the layers. The total depth is h_R, and h_1 and h_2 denote the thicknesses of the upper and lower layers, respectively. The velocities V_1 and V_2 are the average velocities in the corresponding layers, and the shear stresses at the interface and bottom are τ_i and τ_b. The directions of the flows may be the same or opposite or either layer may have an average velocity equal to zero. A one-dimensional equation of motion may be written for the nonuniform flow in each layer; vertical accelerations and side-wall boundary shears are neglected, and only the mean velocities in each layer are considered. The free surface Froude number, based on the total depth and the average velocity of the entire flow, is assumed to be small, compared to unity. Thus, changes in the total depth h_R are neglected, in comparison with changes in the interfacial depth and it is assumed that

$$h_R = h_1 + h_2 = constant \qquad \text{(Equation 1)}$$

The equations of motion and the continuity equation for each layer can be combined into a single differential equation for the slope of the interface (Harleman, 1961),

$$\frac{dh_2}{dx} = \frac{\dfrac{\tau_b - \tau_i}{\rho g h_2} - \dfrac{\tau_i}{\rho g h_1}}{\dfrac{\Delta\rho}{\rho}\left[F_2{}^2 + F_1{}^2 - 1\right]} \qquad \text{(Equation 2)}$$

where F_1 and F_2 are the densimetric Froude numbers of the upper and lower layers

$$F_1 = \frac{V_1}{\sqrt{g\dfrac{\Delta\rho}{\rho}h_1}} \qquad \text{(Equation 3)}$$

$$F_2 = \frac{V_2}{\sqrt{g\dfrac{\Delta\rho}{\rho}h_2}} \qquad \text{(Equation 4)}$$

The bottom and interfacial shear stresses are given by

$$\tau_b = \frac{\rho f_b}{8}|V_2|V_2 \qquad \text{(Equation 5)}$$

$$\tau_i = \frac{\rho f_i}{8}|V_1 - V_2|(V_1 - V_2) \qquad \text{(Equation 6)}$$

where f_b and f_i are dimensionless friction factors for the bottom and interface.

Equation 2 then becomes

$$\frac{dh_2}{dx} = \frac{\dfrac{f_b}{8gh_2}|V_2|V_2 - \dfrac{f_i}{8gh_2}\left(\dfrac{h_R}{h_R - h_2}\right)|V_1 - V_2|(V_1 - V_2)}{\dfrac{\Delta\rho}{\rho}\left[F_2{}^2 + F_1{}^2 - 1\right]} \qquad \text{(Equation 7)}$$

Equation 7 is the general equation for interfacial slope in a two-layer stratified flow.

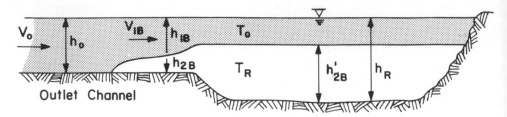

Fig. 3——. Cross-section along center line of outlet channel.

Outlet Channel, Zone B

A cross-section of the river, taken along the center line of the condenser-water outlet channel (section B-B, in Figure 1) is shown in Figure 3. The condenser-water discharge Q_o and its temperature T_o are determined by the power-plant design. Mixing of the condenser-water discharge with the ambient river water at the junction between the discharge channel and the river will be minimized by allowing a tongue of the denser river water to intrude into the discharge channel, as illustrated in Figure 3. In order for a steady-state condition to exist, the average velocity in the lower layer of river water must equal zero. The upper, moving layer of warmer water will induce a flow of the colder river water immediately below the interface, thereby generating a circulation in the lower layer, such that the average velocity is zero. The interface acts as a moving boundary which prevents the formation of a steep vertical velocity gradient and associated mixing by entrainment. Note that, if the cold-water tongue were replaced by a fixed boundary having the same shape as the interface, a steep velocity gradient and entrainment would occur at the junction of the discharge channel and the river.

The equation for the interfacial slope may be written for this special case in which $V_{2B} = 0$. Therefore, $F_{2B} = 0$ and h_R must be taken as the total depth h_o of the outlet channel. Equation 7 for the outlet channel becomes

$$\frac{dh_{2B}}{dx} = \frac{-\dfrac{f_i}{8gh_{2B}}\left[\dfrac{h_o}{h_o-h_{2B}}\right]V_{1B}{}^2}{\dfrac{\Delta\rho}{\rho}\left[F_{1B}{}^2-1\right]} \qquad \text{(Equation 8)}$$

where x is measured in the direction of the upper-layer flow, toward the junction with the river. The development of the river-water tongue requires an upward (positive) slope of the interface in the discharge channel. Since the numerator of Equation 8 is always negative, the denominator must be negative also, therefore $F_{1B} \leqslant 1$. If the width b_o of the discharge channel is constant, the continuity condition requires that

$$V_o h_o = V_{1B} h_{1B} \qquad \text{(Equation 9)}$$

where V_o and h_o are the velocity and depth in the discharge channel upstream of the intruded river water. Equation 4 for the local densimetric Froude number of the upper layer can be written in the following form, after eliminating V_{1B} by means of Equation 9

$$F_{1B} = F_o \left(\frac{h_o}{h_o - h_{2B}} \right)^{3/2} \qquad \text{(Equation 10)}$$

where F_o is the densimetric Froude number of the outlet channel,

$$F_o = \frac{V_o}{\sqrt{g \, \frac{\Delta \rho}{\rho} \, h_o}} \qquad \text{(Equation 11)}$$

Since $F_{1B} \leqslant 1$, it follows, from Equation 10, that $F_o < 1$ and this becomes the design condition for the outlet channel to minimize mixing. Note that F_o is a constant, whereas F_{1B} is a variable along the intrusion length.

At the junction of the discharge channel and the river, the condenser-water undergoes a rapid expansion as it spreads laterally and longitudinally in the river. Thus, a critical depth is established in the upper layer at the junction, such that

$$(F_{1B})_c = 1$$

Note that Equation 8 indicates that, under this condition, the slope of the interface at the junction is vertical. This is the usual condition for a critical depth in one-dimensional channel flow. A similar condition at the junction of a river and an ocean with a negligible tide is known as an arrested saline wedge (Stommel and Farmer, 1953; Harleman, 1961; Keulegan, 1966).

The depth of the upper layer at the junction is given by Equation 10 with $F_{1B} = 1$,

$$\frac{h_o - (h_{2B})_c}{h_o} = F_o^{2/3} \qquad \text{(Equation 12)}$$

In the absence of mixing, the same depth for the heated layer may be assumed to occur in the river adjacent to the outlet channel, thus (see Figure 3).

$$h_R - h'_{2B} = h_o F_o^{2/3} \qquad \text{(Equation 13)}$$

The choice of the densimetric Froude number of the outlet channel is limited by practical considerations. As F_o approaches unity, the tongue of river water will be expelled from the outlet channel, and mixing will occur at the junction. The length of river-water intrusion into the discharge channel increases as F_o decreases. On the basis of a limited amount of experimental evidence (Wigh, 1967), a value of $F_o = 0.5$ appears to be desirable. Larger values result in intrusion lengths which are too small to be effective in reducing mixing and smaller values lead to unnecessarily long discharge channels. The condenser-water discharge rate Q_o and the density ratio $\Delta\rho/\rho$ are fixed by the power plant design; therefore, the remaining design variables are the width b_o and depth h_o of the discharge channel. Setting $F_o = 0.5$ and

$$Q_o = b_o h_o V_o \qquad \text{(Equation 14)}$$

Equation 11 can be expressed as

$$\frac{Q_o}{0.5 \, b_o \sqrt{g \dfrac{\Delta\rho}{\rho}}} = h_o^{3/2} \qquad \text{(Equation 15)}$$

The larger the value of b_o, the smaller will be the depth of the heated layer in the river. For example, if $Q_o = 1000 \, cfs$ and $\Delta\rho/\rho = 0.002 \, (\Delta T = 12° \, F)$ the following combinations of width and depth satisfy Equation 15:

b_o (ft)	h_o (ft)	$h_R - h'_{2B}$ (ft)
250	10	6.3
100	19	12

The depth of the heated layer in the river, $h_R - h'_{2B}$ is found from Equation 13 with $F_o = 0.5$.

Experimental results obtained in a laboratory channel having the geometry of Figure 1 are shown in Figure 4 (Wigh, 1967).

The depth of the warm-water layer was measured in the center of the channel opposite the condenser-water outlet, and the agreement with the theory is good.

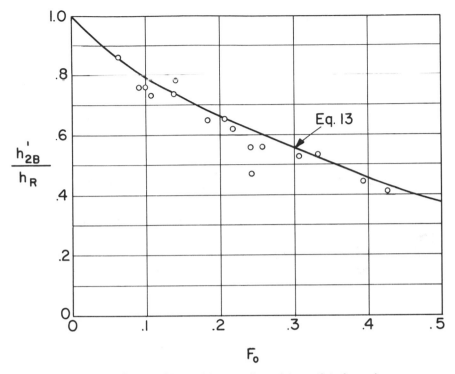

Fig. 4——. Depth of heated layer adjacent to outlet channel.

Middle Region, Between Intake and Outlet, Zone C

The interface between the heated upper layer and the ambient river water in the middle zone of the waterway (Zone C, Figure 1), between the intake and outlet, is considered in this section. The interfacial-slope equation can be written for the special case in which the mean velocity in the upper layer is zero, thus $V'_1 = 0$ and $F'_1 = 0$. This assumes that there is no recirculation of the heated condenser-water layer at the intake. Equation 7 with x measured in the downstream direction, becomes

$$\frac{dh'_2}{dx} = -\frac{\dfrac{(V'_2)^2}{8gh'_2}\left[f'_b + f'_i\left(\dfrac{h_R}{h_R - h'_2}\right)\right]}{\dfrac{\Delta\rho}{\rho}\left[(F'_2)^2 - 1\right]} \qquad \text{(Equation 16)}$$

Equation 16 can be integrated to find h'_2 as a function of x, starting with the known interface height at the outlet. In general, $F'_2 < 1$ and Equation 16 indicates a negative slope; thus, the elevation of the interface at the

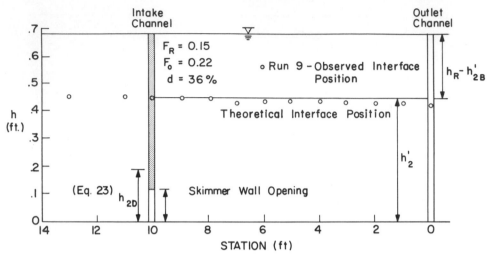

Fig. 5——. Interface in river between intake and outlet channels.

intake tends to be slightly higher than at the outlet. Unless the intake and outlet are separated by a very large distance, the change in the interfacial height will be small. As a first approximation, it is sufficient to assume a constant thickness for the upper layer. The velocity V_2' is the average velocity in the lower layer

$$V_2' = \frac{Q_R - Q_o}{b_R h_2'} \qquad \text{(Equation 17)}$$

where Q_R = river discharge
 Q_o = condenser-water discharge
 b_R = average width of river

Wigh (1967) has shown that there should be no intrusion of the heated water in the middle region upstream of the outlet if the Froude number

$$\frac{Q_R - Q_o}{b_R \sqrt{g \frac{\Delta\rho}{\rho} h_R^3}} > 1 \qquad \text{(Equation 18)}$$

In practice, this value need only be greater than 0.7 to be effective in preventing upstream intrusion of the warm water in the middle region.

Figure 5 shows a profile of the interface on the middle zone, as measured in the laboratory channel. The thickness of the upper layer is essentially constant and equal to the value at the outlet predicted by Equation 13.

SKIMMER WALL

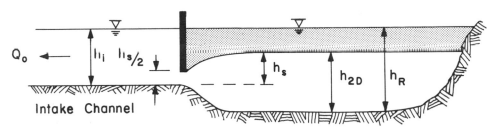

Fig. 6——. Cross-section along center line of intake channel.

Skimmer Wall and Intake Channel, Zone D

When the heated condenser-water layer extends upstream to the intake, recirculation can be controlled by means of a skimmer wall. A cross-section of the river taken along the center line of the intake channel (section D-D, in Figure 1) is shown in Figure 6. The theoretical relationships for the skimmer wall and their experimental verification have been given in an earlier paper (Harleman and Elder, 1965). The design quantities for the skimmer wall are the condenser-water discharge rate Q_o, the density ratio due to the temperature differential $\frac{\Delta\rho}{\rho}$, the width and depth of the skimmer-wall intake channel b_i and h_i, and the head h_s as shown in Figure 6. These quantities are related by the following equation:

$$Q_o = b_1 \sqrt{g \frac{\Delta\rho}{\rho}\left(\frac{2}{3} h_s\right)^3} \qquad \text{(Equation 19)}$$

Equation 19 is obtained from the condition that the local densimetric Froude number is equal to unity at the skimmer wall. Under this condition, the interface will be drawn down to the gate lip. If the flow exceeds that given by Equation 19, the interface will be depressed below the gate lip, and recirculation will occur, since a portion of the intake flow will come from the heated upper layer. In theory, the skimmer-wall opening should be less than $(2/3) h_s$; however, experiments indicate that the opening should not be greater than $(1/2) h_s$. From the geometry of Figure 6, a relation between h_s and h_i is obtained

$$h_s = h_i - (h_R - h_{2D}) \qquad \text{(Equation 20)}$$

where h_{2D} is the depth of the lower layer in the waterway adjacent to the skimmer wall.

The remaining problem arises from the fact that there may be an abrupt depth change in the lower layer of the waterway at the skimmer wall. Two possibilities exist, in the notation of Figure 1,

(a) $h_{2D} = h'_{2D}$

or

(b) $h_{2D} < h'_{2D}$

where case (a) is shown by a solid line at the interface and case (b) by the dashed line in region E. Case (a), where the thickness of the upper layer is governed by the outlet channel design, has already been discussed. Case (b) corresponds to the critical or minimum depth for the lower-layer flow in the river, hence the local densimetric Froude number in the river at the upstream side of the intake equals unity

$$\frac{V_{2D}}{\sqrt{g \frac{\Delta \rho}{\rho} h_{2D}}} = 1 \qquad \text{(Equation 21)}$$

The continuity equation in region E requires that

$$V_R h_R = V_{2D} h_{2D} \qquad \text{(Equation 22)}$$

and Equation 21 may be written as

$$\frac{(h_{2D})_c}{h_R} = F_R^{2/3} \qquad \text{(Equation 23)}$$

where F_R is the densimetric Froude number of the river, defined as

$$F_R = \frac{V_R}{\sqrt{g \frac{\Delta \rho}{\rho} h_R}} \qquad \text{(Equation 24)}$$

In a given situation, the determination of whether case (a) or case (b) prevails depends on the magnitude of F_R and the diversion ratio d, where

$$d = \frac{Q_o}{Q_R} \qquad \text{(Equation 25)}$$

This question is resolved by writing a momentum equation for the lower-layer flow in the waterway between sections upstream and downstream of

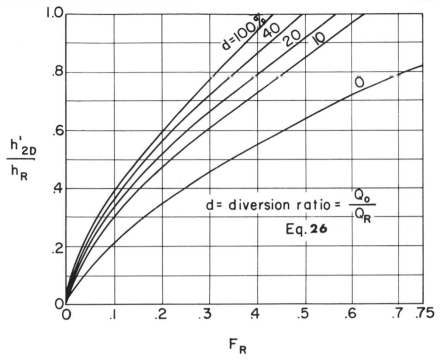

Fig. 7——. Determination of interface depth adjacent to intake channel.

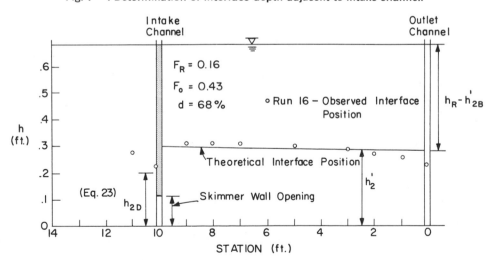

Fig. 8——. Interface in river between intake and outlet channels.

the intake channel. This represents an internal hydraulic jump for the condition in which the discharge per unit-width downstream of the intake channel is reduced by the amount of water diverted to the power plant. The resulting equation (Wigh, 1967) is

$$d = 1 - \frac{1}{F_R}\sqrt{\frac{1}{2}\frac{h'_{2D}}{h_R}\left[3F_R^{4/3} - \left(\frac{h'_{2D}}{h_R}\right)^2\right]} \quad \text{(Equation 26)}$$

and a plot of the family of curves is shown in Figure 7. The ratio $\dfrac{h'_{2D}}{h_R}$

is given by Equation 13 and F_R by Equation 24. With these two quantities, a value of d can be found from Figure 7. If the actual diversion ratio for the prototype is less than the value of d given by Figure 7, the internal jump cannot occur and $h_{2D} = h'_{2D}$ (case a). If the actual diversion ratio is equal to or greater than the value from Figure 7, the jump will occur and h_{2D} will be given by Equation 23 (case b). In either case, h_s is obtained from Equation 20 and the skimmer-wall design-width from Equation 19. Figure 8 shows a profile of the interface in the middle zone under conditions in which case (b) occurs. The profile in Figure 5 represents case (a).

The following example is based on an outlet-channel design-width of 250 ft.

$Q_R = 2500\ cfs$	$b_R = 400\ ft$	$h_R = 25\ ft$
$Q_o = 1000\ cfs$	$\Delta\rho/\rho = 0.002$	$h_i = 20\ ft$
$d = 0.40$	$h'_{2D}/h_R = 0.75$	$F_R = 0.20$

From Figure 7, $d > 1$, therefore $h_{2D} = h'_{2D}$ and from Equation 20, $h_s = 13.7\ ft$. The intake channel width determined from Equation 19 is $b_i = 145$ ft. and the skimmer-wall opening should be not more than 7 ft.

Summary

The foregoing sections have demonstrated the interrelationship between the design of the outlet and intake structures for stratification of the condenser water in the adjacent waterway. The primary requirement for stratification of the condenser-water discharge in the waterway is an outlet channel designed to have a densimetric Froude number less than unity. Heat dissipation in the region between the outlet and intake has been neglected. The calculation of thermal effects in the region downstream of the outlet (zone A) and upstream of the intake (zone E) requires consideration of surface-heat dissipation and is beyond the scope of the present study. If the densimetric Froude number F_R in Equation 24 is greater than unity, the warm-water wedge will not extend into region E upstream of the intake.

The author is indebted to Bata (1957), whose pioneering studies provided much of the background for these developments.

COMPLETE MIXING OF CONDENSER-WATER DISCHARGE

The heated condenser water can be mixed into the natural flow of the receiving waterway by means of a diffuser pipe extending across the river bottom, transverse to the direction of flow. The dimensions of the diffuser pipe and the holes on the perimeter can be chosen to produce a uniform discharge per unit length. The jet of heated water produced by each hole is characterized by a densimetric Froude number of the form

$$F_j = \frac{V_i}{\sqrt{g \left(\frac{\Delta\rho}{\rho}\right)_j D}} \qquad \text{(Equation 27)}$$

where
   $V_j = $ *velocity of diffuser jet*

$\left(\dfrac{\Delta\rho}{\rho}\right)_j = $ *density difference between condenser water and ambient river water*

   $D = $ *jet diameter*

For a prototype condition in which $V_j = 10$ *ft/sec*, $D = 2$ *in.* and the temperature differential is 25° F., the magnitude of F_j is approximately 50. This will result in a high degree of entrainment and essentially complete mixing with the ambient flow. This method of condenser-water discharge produces the minimum temperature rise in the receiving waterway; however, the heat storage in the waterway is a maximum, because heat dissipation to the atmosphere is reduced by the lower water-surface temperatures. Thermal stratification downstream of the diffuser is eliminated; however, a heated layer may develop upstream of the diffuser and cause recirculation.

If the diffuser is assumed to produce complete mixing of the condenser water with the river flow passing over the diffuser, an expression for the downstream mixed temperature, T_M, can be obtained from a one-dimensional thermal-energy balance. Figure 9 shows a control volume whose lateral boundaries coincide with the width of the river equal to the length

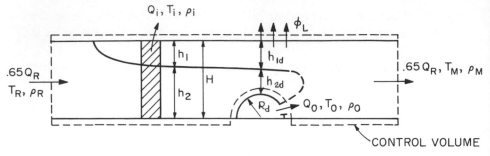

Fig. 9——. Heat balance.

of the transverse diffuser pipe. The control volume extends upstream, beyond the condenser-water intake and the top of the warm-water wedge, and downstream to the region of complete mixing. The heat-flux terms are of the form $\rho c Q T$ (BTU/sec); however, variations in the specific heat c and density ρ may be neglected in the summation of heat fluxes. The energy balance may be written as

$$Q_R T_R + Q_o T_o = Q_i T_i + Q_R T_M + \phi_L \quad \text{(Equation 28)}$$

The various quantities are defined in Figure 9, and ϕ_L is the net rate of heat dissipation from the water surface. The intake flow-rate Q_i is equal to the condenser-water discharge Q_o, hence

$$Q_i = Q_o \qquad \text{(Equation 29)}$$

and

$$T_o = T_i + \Delta T \qquad \text{(Equation 30)}$$

therefore, if the heat-dissipation term is neglected, Equation 28 becomes

$$T_M = T_R + \frac{Q_o}{Q_R} \Delta T \qquad \text{(Equation 31)}$$

Note that the downstream mixed temperature given by Equation 31 is independent of the amount of recirculation at the intake. As the recirculation increases, the discharge of the ambient river water over the diffuser also increases.

The conditions for the existence of the thermal wedge upstream of the diffuser and its thickness at the diffuser can be obtained in the following manner. Figure 9 shows the definition of the wedge thickness h_{1d} at the diffuser. It is postulated that the equilibrium position of the wedge will occur when the depth of the lower-layer flow over the diffuser is critical. Thus, the local densimetric Froude must be equal to unity,

$$F_{2d} = \frac{V_{2d}}{\sqrt{g \left(\frac{\Delta \rho}{\rho} \right)_M h_{2d}}} = 1 \qquad \text{(Equation 32)}$$

The average velocity of the lower layer over the diffuser is

$$V_{2d} = \frac{Q_R - Q_o}{L_d h_{2d}} \qquad \text{(Equation 33)}$$

where L_d is the transverse length of the diffuser.

Combining Equations 32 and 33, it follows that

$$h_{2d} = \left[\frac{Q_R - Q_o}{L_d \sqrt{g \left(\frac{\Delta \rho}{\rho} \right)_M}} \right]^{2/3} \qquad \text{(Equation 34)}$$

In the above equations $\left(\frac{\Delta \rho}{\rho} \right)_M$ is the density ratio corresponding to the

temperature difference $T_R - T_M$. From the geometry of Figure 9, the depth of the thermal wedge at the diffuser is given by

$$h_{1d} = h_R - (h_{2d} + R_d) \qquad \text{(Equation 35)}$$

where R_d is the elevation or radius of the diffuser pipe above the river bed. If $h_{1d} = 0$, the thermal wedge will not penetrate upstream of the diffuser. The river discharge for this condition is found by setting the right side of Equation 34 equal to $h_R - R_d$. For smaller river discharges, recirculation can be controlled by means of a skimmer wall at the intake.

Browns Ferry Diffuser

The Tennessee Valley Authority is constructing a 3300-mw nuclear-power plant at Browns Ferry, 20 miles upstream from Wheeler Dam on the Tennessee River. The condenser-cooling water is to be discharged

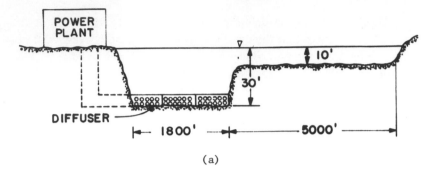

(a)

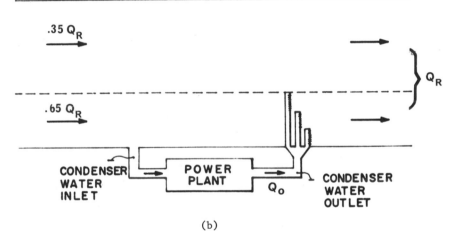

(b)

Fig. 10——. River section at power plant location.

through a manifold of diffuser pipes traversing the bottom of the main river channel. A schematic diagram of the site is shown in Figure 10. Prototype measurements indicate that approximately 65 percent of the total river discharge is through the main channel, which is 30 feet deep and 1800 feet wide. The diffusers are designed to discharge the condenser water uniformly across the width of the channel. The temperature increase through the power plant is approximately 25° F. The diffuser consists of three corrugated steel pipes, each 20 feet in diameter, with the lower half buried in the river bed. The total condenser-water discharge is 4350 cfs and each pipe discharges one-third of this amount into a 600-foot portion of the channel. The jet ports in the diffuser pipe are two inches in diam-

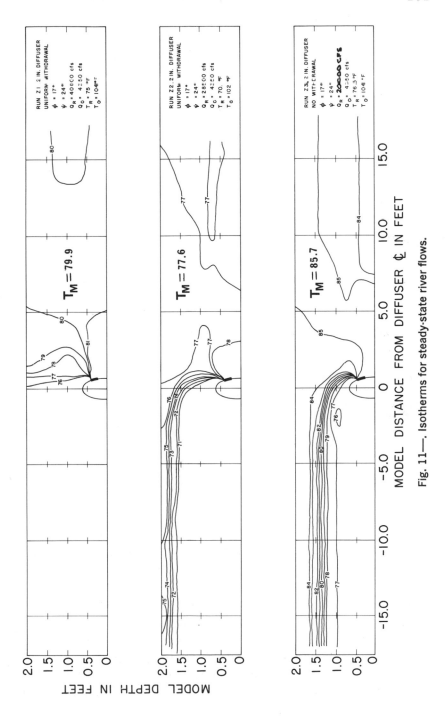

Fig. 11—. Isotherms for steady-state river flows.

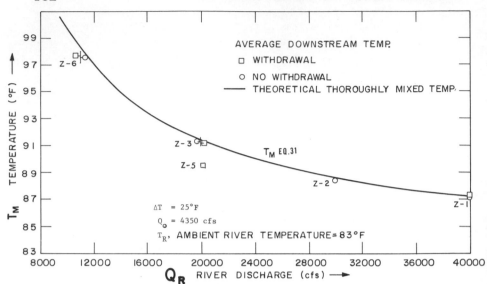

Fig. 12——. Downstream temperature.

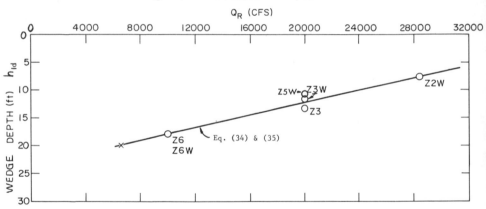

Fig. 13——. Critical wedge depth at the diffuser.

eter, and they are arranged in alternating vertical columns of six and seven ports, spaced six inches on center. The jet ports are located on the downstream quadrant of the diffuser; the first horizontal row of ports is at an angle of 24° from the river bed, and the top row is at an angle of 41°. The former angle was chosen to minimize scour, downstream of the diffuser pipe.

A two-dimensional, undistorted model of a section of the diffuser pipe was tested in the Hydrodynamics Laboratory (Harleman, Hall and Curtis, 1968). The scale ratio was 1/15, and both free-surface and densimetric

Froude similitude was achieved. Plots of the isotherms, upstream and downstream of the diffuser, are shown in Figure 11 for three tests with total river discharges ranging from 20,000 to 40,000 cfs. In each case, essentially complete mixing is obtained in the downstream direction within a prototype distance of 100 to 150 feet from the diffuser. The theoretical mixed temperature T_M is also shown on each plot; in general, the measured temperatures are within 0.5° F. of the values given by Equation 31. In this study, the value of Q_R in Equation 31 is 65 percent of the total river discharge. It can be seen that there is no upstream wedge when the total river discharge is 40,000 cfs. For the lower discharges, the wedge is present and its thickness increases as the discharge is reduced.

Figure 12 shows a comparison of the measured and predicted mixed temperature, as a function of the total river discharge, for an assumed prototype maximum ambient river temperature of 83° F. The measured and predicted thermal-wedge thickness at the diffuser is shown in Figure 13. The experimental program also included a study of the effect of unsteady river discharge on the temperature distribution in the vicinity of the diffuser.

CONCLUSIONS

It has been shown that analytical techniques based on the mechanics of thermally stratified flow can be used in the design of condenser-water intake and discharge structures. The designer has a high degree of flexibility in controlling the environmental changes in a waterway receiving thermal discharges from a power plant.

Thermal conditions ranging from stratification of the effluent to complete mixing can be achieved through proper design. Additional research remains to be done, particularly in the zones upstream of the intake and downstream of the outlet, where heat dissipation must be considered. The effects caused by three-dimensional geometry will probably require thermal model studies in many instances. Recent investigations by the author and his colleagues on thermal modeling are given in Harleman and Stolzenbach (1967 and 1968).

REFERENCES

Bata, G. L. 1957. "Recirculation of Cooling-Water in Rivers and Canals." *Proc. ASCE, Journal of the Hydraulics Division* 83(HY3).

Harleman, D. R. F. 1961. "Stratified Flow." In *Handbook of Fluid Dynamics,* edited by V. L. Streeter. New York: McGraw-Hill.

Harleman, D. R. F., and R. A. Elder. 1965. "Withdrawal from Two-Layer Stratified Flows." *ASCE, Journal of the Hydraulics Division* 91(HY4):43–58.

Harleman, D. R. F., L. C. Hall, and T. G. Curtis. 1968. "Thermal Diffusion of Condenser Water in a River During Steady and Unsteady Flows with Application to the TVA Browns Ferry Nuclear-Power Plant." *MIT Hydrodynamics Laboratory Technical Report No. 111.*

Harleman, D. R. F., and K. D. Stolzenbach. 1968. "A Model Study of Proposed Condenser-Water Discharge Configurations for a Nuclear-Power Plant at Plymouth, Massachusetts." *MIT Hydrodynamics Laboratory Technical Report No. 113.*

Harleman, D. R. F., and K. D. Stolzenbach. 1967. "A Model Study of Thermal Stratification Produced by Condenser-Water Discharge." *MIT Hydrodynamics Laboratory Technical Report No. 107.*

Keulegan, G. H. 1966. "The Mechanics of an Arrested Saline Wedge." In *Estuary and Coastline Hydrodynamics,* edited by A. T. Ippen. New York: McGraw-Hill.

Stommel, H., and H. G. Farmer. 1953. "Control of Salinity in an Estuary by a Transition." *Journal of Marine Research* 12(Jan.):13–20.

Wigh, R. J. 1967. "The Effect of Outlet and Intake Design on Cooling Water Recirculation." S. M. thesis, MIT Department of Civil Engineering.

DISCUSSION/ Norman H. Brooks

DR. HARLEMAN has given a very fine and lucid paper, illustrating that the type of hydraulic structure used for intake and discharge of cooling water has an important effect on the temperature-distribution of the water environment. It is also important, for the power-plant operation, to avoid direct recirculation of heated water, by proper arrangement of intake and outlet. In the hydraulic design of the outlet structures, there are two extremes which can be considered: 1) surface-spreading of hot water with minimal mixing; and 2) extensive jet-mixing of the effluent with the receiving water.

On the one hand, with surface-spreading, the rate of heat transfer to the atmosphere is maximized because the interfacial temperature difference is as large as possible. By avoiding mixing, the volume of water which has a higher-than-normal temperature is minimized. Recirculation to the intake is more easily avoided. On the other hand, with extensive mixing of the hot-water discharge into a reservoir flow or an ocean current, the temperature rise above the background level is minimized, although the volume of water affected by a temperature rise is greatly increased. Furthermore, the rate of heat loss from the water system is greatly reduced, because the temperature differential at the air-water interface is much reduced.

The specification of only the maximum allowable temperature rise (Δt) or maximum temperature is probably not sufficient for regulating thermal-pollution effects. For example, there may be instances in lakes where the ecological impact of a small surface-warm area is less than that for widespread diffusion of heat throughout an entire lake. This effect may be especially important if the wide dispersal and storage of heat delays the fall overturn and accelerates the spring stratification; with a longer interval of stratification, the bottom annual overturn may be incomplete or may not occur at all, if too much diffused heat is added to a lake.

Thus, it may be said that, when mixing devices are used to keep Δt

within allowable limits, there is likely to be a corresponding reduction in the heat flux from the water to the atmosphere by all processes: radiation, evaporation, and conduction. However, in the ocean, which is a tremendous heat-sink, the problem is, mainly, to disperse the heat from the point of discharge, rather than the ultimate exchange of heat from the ocean to the atmosphere or space. Thus, special diffusion structures can be very effective in the ocean or in large reservoirs where the *over-all* heat-balance of the water resource is not the main problem.

If multiple-jet diffusers are to be used for mixing hot-water discharges with the receiving water, then the designer may make use of some of the experience and technical literature on mixing sewage discharges into oceans, estuaries, and lakes (Abraham, 1967; Fan and Brooks, 1966; Brooks and Koh, 1965; Cederwall, 1968; Rawn, Bowerman, and Brooks, 1961; Fan, 1967). Historically, the term *sewage outfall* implied dumping of waste water at the shore. Because this was unsatisfactory, the *submarine* outfall evolved, as soon as underwater construction methods were developed. Effluents could then be discharged from the open end of an outfall pipe, at some distance from shore, at the bottom of the ocean or estuary. In the ocean, the freshwater effluent would rise rapidly to the surface as a large buoyant jet, with modest entrainment of seawater, and then spread over the ocean surface as a buoyant patch. This is analogous to the discharge of hot water from a submerged open pipe, as it, too, would rise to the surface and then spread laterally.

The next evolutionary step, which became common in the 1950s, was the use of large multiple-jet diffusion pipes to distribute the effluent over a larger distance in the ocean and to get much more rapid jet-mixing of the effluent with seawater. Many small jets are used, instead of one large one.

A further innovation in waste disposal is the use of the thermal- (or density-) stratification of the ocean as a barrier to prevent completely the surfacing of waste-water clouds. With many small jets, the buoyant effluent can be mixed with enough colder, bottom water to make the entire mixture heavier than the overlying warm water. This is shown schematically in Figure 1.

It is possible to achieve the same effect with a cooling-water discharge; hot-water jets can be mixed with enough cold, bottom water in a large lake or the ocean so that the buoyant jets are completely trapped below the thermocline. For example, if cooling water at 25° C. is mixed with 10 parts of seawater at 10° C., the mixture-temperature will only be 11.4° C. If the surface layer is warmer than this, then the mixed cloud will be

SUBMARINE OUTFALL WITH
MULTIPLE-PORT DIFFUSER

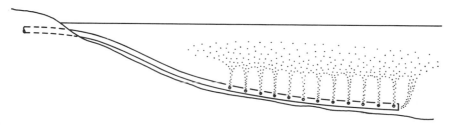

Fig. 1——. Submarine outfall with multiple-port diffuser. Density stratification may keep sewage effluents or thermal discharges below the surface when many small jets are used.

PROFILE

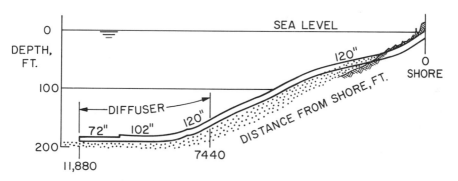

Fig. 2——. Sewage effluent outfall at Whites Point, County Sanitation Districts of Los Angeles.

stopped by the thermocline and will spread laterally, underneath the thermocline. Thus, it would be possible to store waste-heat in a large reservoir by thickening the epilimnion with the ultimate release of heat being primarily during and after the fall overturn.

A schematic of a recently built ocean outfall is shown in Figure 2. This is the 10-foot diameter outfall of the County Sanitation Districts of Los Angeles, at Whites Point, built in 1965 for discharge of sewage effluent following primary treatment. The diffuser section is 4,440 feet long, with

742 ports, ranging from 2.0 inches to 3.6 inches in diameter. It does produce a submerged cloud of diluted sewage effluent most of the year; this reaches the surface only when the amount of temperature differential between the surface and a 200-foot depth drops below about 2° C. Whether the cloud stays submerged or not, the treated sewage effluent is highly diluted with seawater, approximately 200 to 1. (Ironically, it is the same kind of stable stratification in the atmosphere which inhibits the dispersal of the Los Angeles smog layer.)

Jet-mixing produced by a diffuser is just the first stage of dilution. The final stage is the natural diffusion that occurs as the cloud is carried away by an ocean current or a lake current. The tendency in the design of ocean sewer-outfalls is to depend largely on the first stage, or man-controlled mixing with large diffusers. The use of good jet diffusion means that the stratification can be beneficially utilized and shorter ocean outfalls will produce satisfactory water quality. Similarly, in thermal-discharge problems, if large mixing is desired, it is probably most feasible to produce the mixing immediately by a multiple-jet diffusion structure, rather than to count on the slower and less dependable natural mixing. Only a few feet of additional hydraulic head are required to operate a large diffuser.

Figures 3 and 4 illustrate, in a laboratory tank, the behavior of buoyant jets discharged horizontally. In Figure 3, the water is of uniform density throughout, whereas, in the case of Figure 4, there is a linear stratification from top to bottom. The densities and the stratifications are shown schematically in Figures 5 and 6, respectively. The limited height of rise in the stratified case (Figure 4) is clearly indicated.

In recent years, the fluid mechanics of buoyant jets has been sufficiently worked out for use in design problems. For round buoyant jets discharged horizontally in a uniform environment, see Abraham (1967) and Fan and Brooks (1966), for example. For the same situation with stratification, see Fan (1967). For a two-dimensional source, such as a row of jets discharging into a stratified environment, see Brooks and Koh (1965). For a comprehensive compilation of various numerical solutions to buoyant-jet problems, see Fan and Brooks (1968). The above-mentioned references presume no ambient current; Fan (1967) has also given an analysis for a buoyant jet in a uniform current without stratification. Space does not permit a more detailed discussion of these analyses here.

When a diffusion structure becomes large, compared to the depth of flow, then the momentum input may affect the over-all flow-pattern. Then it is necessary to make a model study, such as the one described by Dr.

Fig. 3——. Buoyant-jet discharge into a homogeneous fluid in a laboratory tank.

Fig. 4——. Buoyant-jet discharge into a stratified fluid in a laboratory tank (at same discharge as for Fig. 3).

Harleman in his paper. The various buoyant-jet formulas which presume infinite-flow fields simply do not apply.

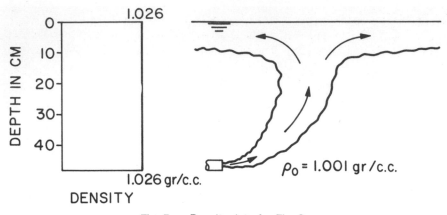

Fig. 5——. Density data for Fig. 3.

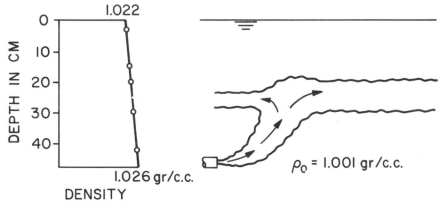

Fig. 6——. Density data for Fig. 4.

To design a diffuser for mixing requires careful consideration of the hydraulics *inside* the diffuser, also. It is essentially a manifold problem, with many lateral discharges for which it is usually desired to have uniform efflux along the length of the diffuser. The procedure for making such a hydraulic analysis is illustrated by Vigander, Elder and Brooks (1968) on the design of the Browns Ferry diffuser pipes (TVA).

In summary, I wish to make the following points:

It is possible to make hydraulic designs for cooling-water discharges, ranging from surface-spreading with minimal mixing to complete mixing with possible storage of waste heat below thermoclines. The alternatives elected will certainly depend on the predicted ecological effects of managing the waste heat in different ways. In general, the strong-mixing approach can be expected to be most useful in the ocean or in reservoirs and estuaries in which there are substantial currents to carry away the waste heat from the plant site. In small bodies of water, or those without substantial currents, surface-spreading will probably be more effective, both for getting rapid transfer to the atmosphere and to avoid recycling.

It is clear that, in the decade ahead, with the rapidly expanding power industry, there simply will not be enough cooling capacity in the inland rivers of the United States. This freshwater part of our hydrosphere is only a few hundredths of one percent of the total water, and it is too valuable for other uses. Therefore, I predict that there will be far greater usage of cooling towers, and much more pressure to locate power plants on the coastline, for use of seawater for cooling. With adequate diffusers and good ocean currents, thermal pollution of the ocean can easily be avoided for some time to come. Furthermore, efforts should be made to develop power-generation methods which are more efficient and produce less waste heat per unit of power generated.

I wish now to make one final plea to research workers in this field. I strongly urge the use of the metric system of units in all calculations involving heat transfer and energy. This is, of course, common practice by meteorologists and oceanographers, as well as by the engineers in the "metric countries." The interactions with the scientists in this area would certainly be facilitated by universal use of the metric units. After all, since we already use kilowatts, instead of horse-power, we ought to be able to use calories instead of BTUs, and temperature in Centigrade instead of Fahrenheit. If we can get used to using the metric units in the beginning, then we shall be spared the painful exercise of converting everything back and forth between English and metric units!

REFERENCES

Abraham, G. 1965. "Horizontal Jets in Stagnant Fluid of Other Density." *ASCE, Proc., J. Hydraulics Division* 91(HY4):139–154.

Brooks, N. H., and R. C. Y. Koh. 1965. "Discharge of Sewage Effluent from a Line Source into a Stratified Ocean." In *International Association for Hydraulic Research*, XI Congress, Paper 2–19. Leningrad.

Cederwall, Klas. 1968. "Hydraulics of Marine Waste-Water Disposal." In *Chalmers Institute of Technology, Hydraulics Division*, Goteborg, Sweden, *Report No. 42*.

Fan, Loh-Nien. 1967. "Turbulent Buoyant Jets into Stratified or Flowing Ambient Fluids." *W. M. Keck Laboratory of Hydraulics and Water Resources Report No. KH–R–15*. California Institute of Technology.

Fan, Loh-Nien, and Norman H. Brooks. 1968. "Numerical Solutions of Buoyant Jet Problems." *W. M. Keck Laboratory of Hydraulics and Water Resources Report No. KH–R–18*. California Institute of Technology.

Fan, Loh-Nien, and Norman H. Brooks. 1966. "Discussion of 'Horizontal Jets in Stagnant Fluid of Other Density,' by G. Abraham." *ASCE, Proc., Journal Hydraulics Division* 92(HY2):423–429.

Rawn, A. M., F. R. Bowerman, and N. H. Brooks. 1961. "Diffusers for Disposal of Sewage in Sea Water." *ASCE, Trans.* 126(3):344–388.

Vigander, Svein, Rex Elder, and Norman H. Brooks. 1968. "Diffuser Design for Cooling-Water Discharge at Browns Ferry Nuclear-Power Plant, TVA." Paper read at Conference of American Society of Civil Engineers, May 1968, at Chattanooga, Tennessee.

DISCUSSION FROM THE FLOOR

Thomas P. Gallagher: I would like to ask **Dr. Harleman** whether the decision to go to the completely-mixed system in the reservoirs was based on the Tennessee criterion of a maximum 10° F. temperature rise and also whether any consideration was given to discharge to the surface?

Donald Harleman: I do not feel qualified to speak for TVA in regard to policy decisions, since I am not a part of TVA. We did the laboratory tests at their request. If someone from TVA would like to speak on that matter, I would be happy to have them.

Milo Churchill: We did consider it, of course, but you must realize that we are talking here of a thermal rise of about 25° to 26° F. When you add 25° to 85°, what do you get? You get too much. It's as simple as that. You cannot put that much hot water on the surface of the reservoir: you simply cook everything there. That's as simple an answer as I can give you: you just can't do it.

Edward Silberman: I thought it might be worth pointing out that one can get a stratified flow, even with some mixing at the outlet. For instance, in a horizontal outlet near the surface, one can mix discharge-canal water with ambient water in the river, and come up with a water of different density which will still stratify and form a similar kind of surface current to the one you were illustrating. Such mixing might be produced by waves or by a physical structure. Would you care to comment on that any further?

Harleman: I was not implying that, in order to have stratification, you have to have no mixing. I am simply talking about one end of the scale in which you minimize the mixing intentionally. As you remember, from the slides of the Browns Ferry diffuser, where we have complete mixing downstream, we had quite complete stratification upstream. This is an

173

example of just what you said, that even though you produce mixing downstream, if the conditions in regard to velocity and depth of the river are right, then you will get an intrusion of that mixed water upstream, with consequent stratification. The whole realm of engineering design between complete stratification or minimizing the mixing and complete mixing is possible. I am not saying that we have all of the design information, but it is possible to achieve any degree of mixing you want, and the controlling variable is the densimetric Froude number of the outlet.

John W. Foerster: I would like to know, Dr. Harleman, how you could predict heat dissipation on the following example, with the formulas you have given: We are working with a power plant's effluent about 25 miles up the Connecticut River from the seacoast. At the present time, stratified-flow conditions exist across the river and downstream from the plant-outlet channel on the ebb tide. At flood tide, there is a damming effect, and this heated water is pushed to the east side of the river and upstream, past the water-intake of the power plant, a total of about one mile from the effluent-channel.

Harleman: I think you would need a blackboard to understand the geometry. I think you are saying that the conditions are not uniform transversely across the section. At one stage of the tide they are; and at the other stage, they are not. This is a three-dimensional-stratification problem. The equations I presented are for vertical changes in the interface, and they would not apply, in this case. I suspect you would have to investigate such problems experimentally, under laboratory conditions. I think it would be rather difficult to determine the answer analytically.

H. J. Gormly: In using these three-dimensional models, I think we can go a certain distance down the river from a plant outfall and get some pretty good results. We have done some modeling, using the Bay Model, and have gotten some good results for the "mixing zone." But have you given thought as to how we can numerically describe where we move out of the mixing region and begin to worry about the surface-heat transfer coefficients they were talking about in the previous papers this morning?

Harleman: I think this is one of the areas where research is definitely needed. I believe that all the speakers' preceding papers were talking about heat dissipation from situations in which there is no vertical stratification. The problems of predicting the heat dissipation and variation of interfaces with initial stratification, to my knowledge, have not been adequately solved. The problem probably has to be approached by a two-dimensional type of diffusion equation, with advection and diffusion, rather than from the standpoint of the two-layer stratified situation, because that doesn't admit any mixing. I think one can handle these problems numerically, if data becomes available; field data is very scarce for the situation where you have heat dissipation from an initially stratified situation. I think it depends on getting some field data and then trying some numerical methods to develop techniques. As far as I know, they really have not been developed. One is forced, at the moment, to do hydraulic modeling. When you get into complex three-dimensional situations, we resort to laboratory investigation. We try to recognize the limitations and the difficulties, but probably, at the present time, they are a way of proceeding.

Francis S. Gartrell: I think that touches on a matter that's of concern to a lot of us: What is a reasonable mixing zone? How would you go about defining a mixing zone, when you purposely try to discharge heated water in a way to avoid mixing?

Vern W. Tenney: I am interested in knowing what percentage of time the waste-cloud that you talked about in your slides is submerged.

Norman H. Brooks: You refer to the ocean disposal?

Tenney: Yes, the southern California ocean, I believe it was, outfall.

Brooks: For San Diego, the cloud is submerged for the entire year, so far, on the low discharges that they have in their new outfall since 1963. For the City of Los Angeles and the County of Los Angeles systems, the submergence occurs about 11 months out of the year. It varies from year to year, depending on the occurrence of winter storms, but the isothermal (or constant density) conditions in the ocean occur only for a very few weeks in the middle of winter. In Puget Sound, in Seattle, where we also designed for submergence, the density situation is different, and submergence occurs for something like 8 months of the year, and the sewage cloud would reach the surface for about 4 months; however, I haven't seen the latest results, but I think that's about what it's doing.

Mack S. Prichard: I am interested in the ecologic side-effects of the thermal plants described. Have you noticed, or have you any record of, sea urchins or have there been any increases in sharks on the Los Angeles outfall?

Brooks: I am not qualified to answer that, as I am not a biologist. We *are* concerned with what the ecological effects are, and Professor Wheeler North at Cal Tech is making extensive studies of the ecological effects of waste-water and slight warming of coastal water on the interactions of sea urchins and the whole food chain.

Walter O. Wunderlich: Dr. Brooks, you outlined briefly the theory of a rising plume by a diagram. Is this theory able to predict the spread of the plume in one direction or in a radial direction?

Brooks: It does predict the spreading of the plume in a radial direction. The analysis is based on the assumption of axial symmetry in the plume, assuming that the curvature is not too sharp so that you can follow it in a one-dimensional flow, and using the similarity assumptions as another jet-flow problem.

Wunderlich: I was mainly interested in whether the plume can also spread toward the shore. You showed spread only out into the ocean, but it would depend upon the angle at which the plume meets its density layer as to how much of the jet is directed back toward the shore.

Brooks: In putting together the analysis for the waste-disposal problem, you really have to fit together different parts of it as best you can. In the ocean, the discussion of buoyant plumes neglected the effect of the current. The current actually does tend to sweep these plumes over in certain ways. Once the initial mixing is over, in this so-called mixing zone, which is relatively close to the diffuser, then the whole cloud travels whichever way the current takes it. I will say, though, that when the ocean is stably stratified, if the cloud is neutrally buoyant at -50 feet, then the closest it can get to shore is where the depth is 50 feet. The temperature contour just bumps into the shore, and that's as close as it gets, except in times of really extreme upwelling, which doesn't occur in these very stable summertime months.

Chapter **6**   Peter Ackers

MODELING OF HEATED-WATER DISCHARGES

IT is impossible to deal comprehensively in this paper with hydraulic modeling laws, although some of the general principles involved will require mention. It is intended to concentrate on some of the more practical aspects of the modeling of buoyant effluents, including a comparison of model and prototype observations in several situations.

A prime requirement is similarity of flow pattern in the ambient fluid in the commonly occurring situation where other forces—for example, tides or river currents—result in general motion of the receiving waters. The basis of model design to obtain similarity for this ambient motion is well known: where gravity and inertia are relevant, a necessary condition is equality of Froude Number, $(V/\sqrt{gy})_{model} = (V/\sqrt{gy})_{proto}$ where V is a typifying stream velocity and y is a measure of depth. But in many problems, there are additional requirements for similarity of ambient conditions —for example, bed friction, salinity, and bed movement. On the other hand, there are some cases where the ambient fluid would be at rest but for the influence of the effluent. Thus, the problem may fall in one or other of the following categories:

1. Discharge into an ambient fluid otherwise at rest. This is a condition

This paper has been presented with the permission of the Director of Hydraulics Research, and the majority of the studies cited have been conducted as part of the program of the Hydraulics Research Station of the [British] Ministry of Technology. Photographs are by the British Ministry of Technology, Hydraulics Research Station, Wallingford, Berks. The author acknowledges the help of many of his colleagues; those with direct responsibility for these ad hoc studies, those who have been engaged in fundamental research, and those who have contributed in other ways. Thanks are also due to the University of Strathclyde, Glasgow, [Scotland], and to the several clients at whose expense the ad hoc studies were made (the [British] Central Electricity Generating Board, the Electricity Supply Board of Ireland, the South of Scotland Electricity Board, the Hong Kong Electric Company).

177

not conducive to rapid heat dissipation, the only fluid motions being generated by the effluent itself through its inertia and buoyancy, and by the associated intake.

2. Discharge into a uni-directional flow. This is a condition in which, with a little common sense in siting, intake and outfall recirculation can be avoided. Nevertheless, this is a situation where interactions with adjacent heat sources or the influence of hot water on the biology of the river may make it necessary to build a model as part of the investigation.

3. Discharge into a tidal stream. This is a condition often conducive to rapid heat dissipation, and yet because of current reversals and eddies, recirculation at some tidal state is probable. This, as well as interactions with other power stations, makes modeling desirable.

Most of this report concerns category 3, as this is a common situation in Britain. Our small catchments rule out major power stations on rivers using direct cooling, but the proximity of the coastline and the existence of deep estuaries with very strong tidal flushing has made it practicable to site a large proportion of the British generating capacity on or near the coastline. The nuclear-generation program has reinforced this trend, both because of the use of direct cooling and the desire to site early stations of the series away from large centers of population.

The Dispersal of Heated Water

Modeling of heated water is complicated by the fact that the effluent may disperse by several different mechanisms, and more than one such mechanism may exist at any given site, though in some cases one of the mechanisms may dominate the dispersal. Therefore, before embarking on a model study in this field we must identify the relevant mechanisms for dispersal in the particular case. The design of the investigation—whether a physical model is needed and appropriate, its type, extent, control system, scales, and interpretation, what field work is needed and the extent of any mathematical analysis—will all depend, first, on a proper recognition and, second, on the representation of the relevant dispersal mechanism. And let it be admitted here and now that it is impossible to design and operate a hydraulic model of a situation in which several types of dispersion occur accurately simulating them all.

The main stages of dispersion are:

A. Turbulent entrainment at the efflux jet. Close to the outfall, the inertia of the jet is important, whilst density differences are not.

B. Buoyant rise of jet, if submerged. The trajectory of the jet is dependent on initial inertia as well as buoyancy force, the mixing being due to entrainment by turbulence at the plume boundary.

C. The convective spread of effluent from such an initial dilution zone over the surface of the receiving water. This process is dependent on the density difference where the convective spread has its origin, e.g., from the boil point of a submerged jet. Whether or not mixing will occur at the interface is dependent on the densimetric Froude number.

D. The mass transport of the effluent by ambient currents, which impose additional velocity vectors on all other stages; these will vary periodically in tidal areas.

E. Diffusion and dispersion due to the turbulence in the ambient fluid. This will result in a general temperature dilution superimposed upon the other processes.

F. Additionally, in the nonsubmerged phases, there is loss of heat by evaporative cooling (and in some circumstances, by conduction into solid surroundings).

Outline of Modeling Principles

The different stages of the process that have just been listed imply certain scale relationships, of which the following notes are just an outline:

A. Jet diffusion
  1. Geometric similarity necessary, as there is no means of vertically exaggerating the turbulent structure of the jet.
  2. Outfall Reynolds number large enough to yield turbulent flow *(Re$_{outfall}$ > 2,500 approx)*. The above will yield similar dilution and jet geometry in this simple case, but it is usual to operate model on the basis of:
  3. Froude Law,

$$\lambda_v = \lambda_d^{1/2} \qquad \text{(Equation 1)}$$

where λ *represents the scale of any quantity*
         v = *velocity*
         d = *representative dimension of outfall.*

This is necessary if the jet itself is gravity dependent, such as the nappe from a freely discharging weir, or if the ambient flows are gravity dependent.

B. Buoyant plume
  1. Geometric similarity
  2. Outfall Reynolds number above critical value
  3. Densimetric Froude Law,

$$\lambda_v = \lambda_d^{1/2}\, \lambda_\Delta^{1/2} \qquad \text{(Equation 2)}$$

  where Δ = *density difference relative to ambient*

$$conditions = \frac{\rho_s - \rho}{\rho}$$

C. Convective spread
  1. Densimetric Froude Law,

$$\lambda_v = \lambda_h^{1/2}\, \lambda_\Delta^{1/2} \qquad \text{(Equation 3)}$$

  where h = *thickness of buoyant layer.*
  2. Densimetric Reynolds number, Re_Δ, to exceed certain limiting value. This will be mentioned again later but vertical exaggeration is often needed to meet this requirement.

D. Mass transport by ambient currents
  1. Froude Law,

$$\lambda_v = \lambda_y^{1/2} \qquad \text{(Equation 4)}$$

  where y = *depth of flow.*
  (This may lapse in those special situations where flow is not gravity dependent.)
  2. Reynolds number to yield turbulent conditions, $Vy/v > 600$ *approx,* where v is the kinematic viscosity of the fluid.
  3. Roughness to give correct head loss,

$$\lambda_C = \lambda_v\, \lambda_x^{1/2}\, \lambda_y^{-1} \qquad \text{(Equation 5)}$$

  where C = *Chezy coefficient*
          x = *plan dimension*
  (This may lapse where flow is not friction dependent.)

E. Ambient turbulence
  1. Full geometric similarity is needed, because the turbulent structure cannot be vertically exaggerated.
  2. Given geometric similarity

$$\lambda_C = 1 \qquad \text{(Equation 6)}$$

yields the appropriate value of dispersion coefficient to give similarity of turbulent dispersion.

F. Surface cooling

Heat loss is proportional to area x time x temperature difference assuming that the same "climate" exists in the model shed as in the prototype. The temperature drop is proportional to mass $\div$ heat loss. Hence,

$$\lambda_{\Delta\theta} = \frac{\lambda_{mass}}{\lambda_x{}^2 \lambda_t \lambda_\theta} \qquad \text{(Equation 7)}$$

But to produce equal temperatures in model and prototype, we also need to produce equal temperature drops.

Using the same fluid, $\lambda_{mass} = \lambda_x{}^2 \lambda_y$, and in a Froude model $\lambda_t = \lambda_x \lambda_y{}^{-\frac{1}{2}}$. Inserting in the above, putting $\lambda_{\Delta\theta} = \lambda_\theta = 1$, we find

$$\lambda_x = \lambda_y{}^{3/2} \qquad \text{(Equation 8)}$$

Thus surface cooling requires vertical exaggeration of $\lambda_y{}^{\frac{1}{2}}$.

Certain compatibilities and incompatibilities emerge from the above. There are good reasons for making the density difference at the outfall the same in the model as in the prototype, i.e.,

$$\lambda_\Delta = 1 \qquad \text{(Equation 9)}$$

This permits phases A and B, jet diffusion and buoyant plume, to be modeled in a geometrically similar model of reasonably large size, without vertical exaggeration. Mass transport by ambient currents (stage D) is compatible with this type of model.

When an estuary or other extensive area has to be represented, vertical exaggeration is usually necessary because the vertical scale acceptable from considerations of minimum Reynolds number, control of tides, and measurement of depths and velocities cannot be accepted for the horizontal scale, because of space limitations. With vertical exaggeration, mass transport can be simulated (stage D) and also convective spread (stage C) with $\lambda_\Delta = 1$. Furthermore, it will also be possible by proper selection of scales to represent surface cooling (stage F). But it will be appreciated that vertical exaggeration, necessary as it is in certain problems, prevents the accurate simulation of lateral dispersion by turbulence (stages A and E).

The question might be asked: "Is hydraulic modeling of heated effluents ever justified in view of these incompatibilities?" A hydraulics research laboratory may seem to have a vested interest in physical modeling, but no responsible research team would choose to investigate by model unless they were convinced of a model's merit, having recognized its limitations. Despite the imperfections of the tool, a physical demonstration of recirculation paths in the complex situation of estuary flow is of undoubted value, for only in the simplest circumstances can computation produce convincing evidence about likely thermal pollution.

An obvious way of circumventing the incompatibilities is to use two models when stages A, B, C, and D are all of importance in a particular case:

1. A model to a large natural scale of an area near the outfall, to represent initial dispersion and the buoyant plume zone and to determine the starting condition for the convective and mass transport phases. This could be a steady-state model, rather than a full tidal one, with boundary conditions taking account of model 2 results.

2. A vertically exaggerated model covering the whole zone of interest with full tidal control. The boundary condition imposed at the outfall should take account of the results of model 1.

But this does not overcome the limitations of modeling stage E dispersion. Fortunately, in many situations, stage E is not the major process.

It must be stressed that the above is no more than a brief and oversimplified outline of model rules, to provide a background to the discussion of model techniques and the comparison of model and prototype that follows. The student will know that detailed information on the several phases of dispersion exists in research literature, and he will appreciate that each of these topics is sufficiently complex to warrant extensive and continuing study. Some relevant research is listed in the bibliography.

Examples of Models and Comparisons with Field Data

Maximum advantage should be taken of any opportunity for field studies for comparison with model results. This is desirable as a follow-up after completion of the construction program and commissioning of the station, but is even more profitable if there is an existing heated effluent close to the proposed outfall. Thermal surveys carried out at various tidal conditions will provide invaluable data for proving—and adjusting—the model, and it is the policy at Wallingford to spend an appreciable proportion of the investigation costs on such surveys.

Severn Estuary Model

A few years ago, the Hydraulics Research Station operated a model for

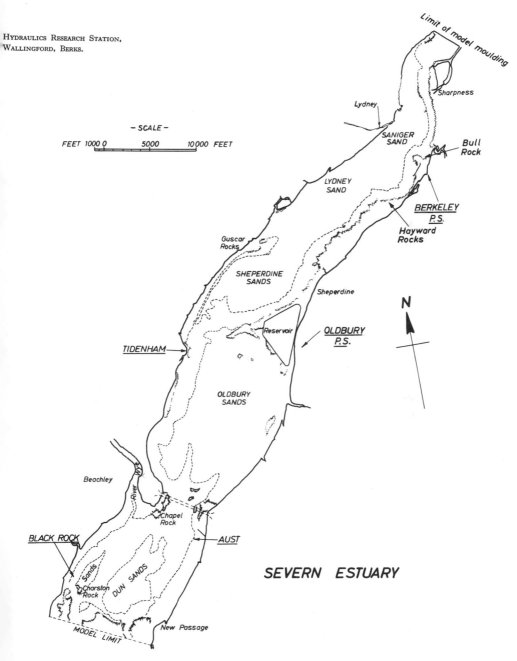

Fig. 1—. Severn Estuary from New Passage to Sharpness. Sites of existing and prospective power stations.

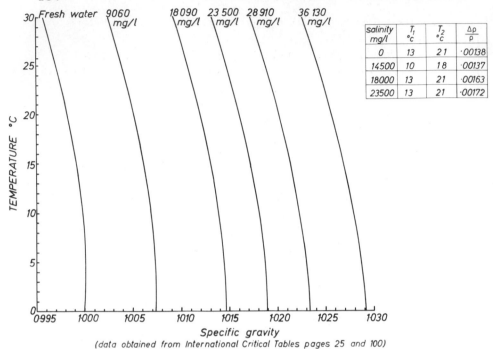

Fig. 2—. Variation of specific gravity with temperature and salinity.

SOURCE: Data obtained from International Critical Tables, pp. 25 and 100.

the Planning Department of the Central Electricity Generating Board to study proposals for the siting of several nuclear-power stations on the Severn Estuary. (Berkeley, 300 megawatts, 1000 cusecs, now in operation; Oldbury, 550 megawatts, 1050 cusecs, under construction; Black Rock, Tidenham, and Aust, each approximately 2000 megawatts, 3000 cusecs, prospective sites; see Figure 1.) The scales of 1/480 horizontal and 1/60 vertical were selected largely on the basis of tidal reproduction and potential changes of bed configuration, it being considered in this area of high tidal range and currents (spring tide range, 30 feet at Berkeley; maximum currents approx 6 ft/sec) that mass transport was the dominant process. The estuary is not stratified, and the time of retention within the cooling systems of the stations is short enough for there to be no significant difference (due to time-dependent longitudinal salinity gradient) between the salinity of the rejected cooling water and the water into which it is discharged. Thus, salinity was not represented in this model. It should be remembered, though, that the coefficient of thermal expansion of water depends on salinity, so that a unit scale of density difference may necessitate some small departure from a unit temperature scale. This may be deduced (Jaffrey *et al.,* in press) from Figure 2, but it may be observed that a difference in background temperature can counteract the omission of salinity: a model operated with fresh water at an ambient temperature of 13° C. (55° F.) will have much the same value of $\Delta\rho/\rho$ for a rise of 8° C. (15° F.) as a prototype with water of salinity 14,500 mg/1 at an ambient temperature of 10° C. (50° F.).

Although the cooling system at Berkeley nuclear station was not one of those studied in detail in the model (design of this station pre-dated our model), observations were made in 1963 of the spread of hot water from Berkeley, as part of the study of possible interactions between adjacent stations. It had been hoped to obtain corresponding data from the site when the station was on full load in 1966, but weather conditions aborted the effort, so field results were not forthcoming until 1967; but the comparison is of interest.

The survey was made at seven distinct tidal conditions, but it suffices here to give details of only one such comparison, for 3 hours before low water of a mean tide. Heat rejection was a shade higher and river flows lower than under the conditions observed in the model, but a correction was made for the former. Figures 3 and 4 show the field measurements (Jaffrey *et al.,* in press) of surface temperature (6 to 9 inches below water surface) and indications are that the path of the plume and the rate of drop in temperature on the axis remote from the outfall were well reproduced in the model. It will be seen from Figure 3 that the effluent is discharged behind a baffle wall into a channel to which tidal currents have access, though with some restriction at the east end by a rock outcrop at late stages of the ebb. For the particular condition illustrated, there is some potential dilution of the effluent behind the baffle wall because of the flow of tidal water, but there is a further zone of shear flow that must involve turbulent dispersion where the flow emerges from the west end of the baffle wall at the edge of the strong tidal stream. We anticipate disparity in a model with a vertical exaggeration of 8 in such a zone, and the evidence is of too great a dilution at the axis here coupled with a too rapid expansion in plan. The comparison for 1 hour before L.W. of a mean tide (prototype data being obtained on two occasions with different wind conditions) showed the same type of divergence, including the tendency for the zone of rapid temperature reduction on the axis to be nearer the end of the baffle wall in the model.

From what has already been said, it will have been realized that, by using a unit scale of temperature, densimetric Froude number similarity was

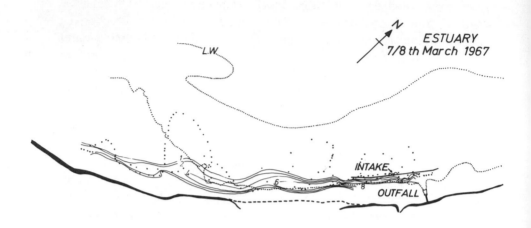

3 HOURS BEFORE L.W. MEAN TIDE
Isotherm values in deg. C. above ambient

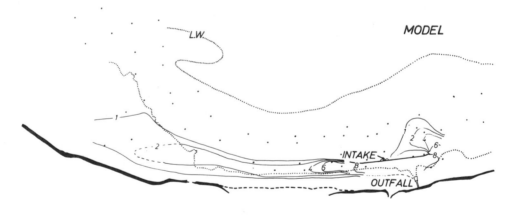

Fig. 3—. Temperature survey at Berkeley Power Station. Comparison of field and
model results.

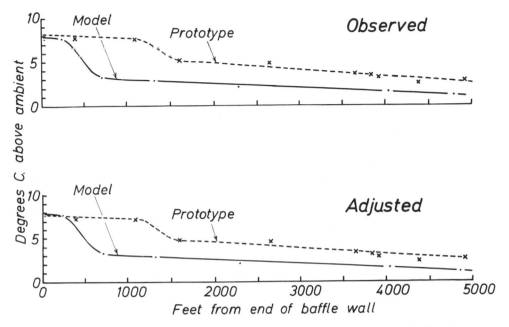

Fig. 4——. Surface temperature on axis of outfall plume three hours before l.w. mean tide. Longitudinal temperature profile at Berkeley.

achieved, but it is instructive to check whether the scales chosen in 1958 satisfied the requirements for one-dimensional convective spread that have emerged from subsequent work on the lock-exchange problem by Barr and his colleagues at the University of Strathclyde.

Figure 5 shows the spread of a buoyant layer under lock-exchange conditions (Frazer *et al.*, 1968). In the right-hand zone, the phenomenon is independent of viscosity, the experimental data confirming the expected relationship

$$\frac{L/T}{\sqrt{(gH\ \Delta\rho/\rho)}} = constant \qquad \text{(Equation 10)}$$

where $L = distance\ the\ layer\ has\ spread\ to$
$T = time\ elapsed$
$H = depth\ of\ fluid$

Thus with $\Delta\rho/\rho$ the same in model and prototype, a buoyant layer will spread with Froudian velocity provided both model and prototype lie in this "non-viscous" zone. For model spread up to a given distance L to lie in this zone, certain scale relationships are implied as follows:

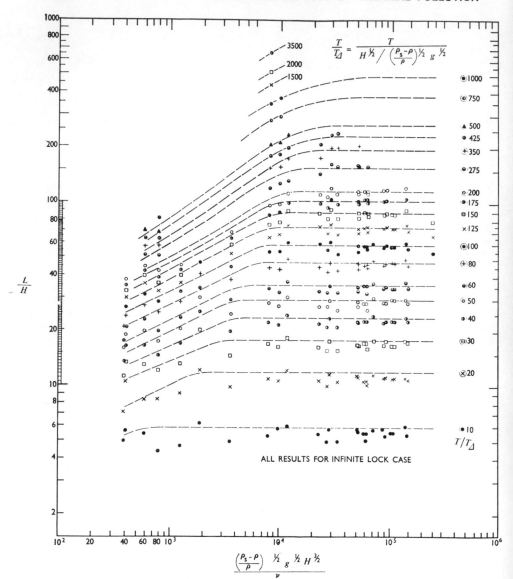

Fig. 5——. Spread of a buoyant layer.

SOURCE: After Frazer **et al.,** 1968

The limit to the non-viscous zone is represented by the equation:

$$\frac{\sqrt{(g\Delta\rho/\rho)}\ H^{5/2}}{Lv} > 150 \qquad \text{(Equation 11)}$$

(This is a dimensionless group derived in 1934 by O'Brien and Cherno.)
Now   (Equation 12)

$$\left[\frac{\sqrt{(g\Delta\rho/\rho)}\ H^{5/2}}{Lv}\right]_{model} = \left[\frac{\sqrt{(g\Delta\rho/\rho)}\ H^{5/2}}{Lv}\right]_{prototype} \lambda_H^{5/2}\ \lambda_L^{-1}\ \lambda_v^{-1}\ \lambda_\Delta^{1/2}$$

When the same density difference and fluid are used, $\lambda_\Delta = 1$ and $\lambda_v = 1$.
Calling the exaggeration ε, where $\varepsilon = \lambda_H/\lambda_L$, the above limit becomes

$$\varepsilon^{5/2}\ \lambda_L^{3/2} > \frac{150}{\left[\dfrac{\sqrt{(g\Delta\rho/\rho)}\ H^{5/2}}{Lv}\right]_{prototype}} \qquad \text{(Equation 13)}$$

This equation can be used to deduce the minimum vertical exaggeration that satisfies the convective spread criterion, although there may be good reasons for using a lesser exaggeration after checking from Figure 5 that the error involved in convective spread to some critical position is small.

In the case of the Severn model, there should be no error in the extension of the front by convection up to $L \cong 600$ feet, 4 percent error at 2,500 feet, and 10 percent at 4000 feet from the outfall. Remembering that ambient currents in this case carry the layer far further than does convection, any disparity in surface isotherms between model and prototype is unlikely to be due to the convective element of the dispersion. The vertical exaggeration selected was also correct for simulation of surface cooling.

A total of 13 reports was prepared by the Hydraulics Research Station in the period 1959 to 1967 on hydraulic and thermal aspects of the siting of power stations on the Severn. In addition to the main estuary model, three other models of local areas were built. Two of these were to a natural scale of 1/60, for the purpose of studying the possibility of local recirculation.

Black Rock Outfall/Intake Model

A tower structure combining a circular outfall weir with a low-level intake was proposed for the Black Rock Station, and it was clear that a plunging nappe, particularly when tide levels were low, might possibly give high local recirculation.

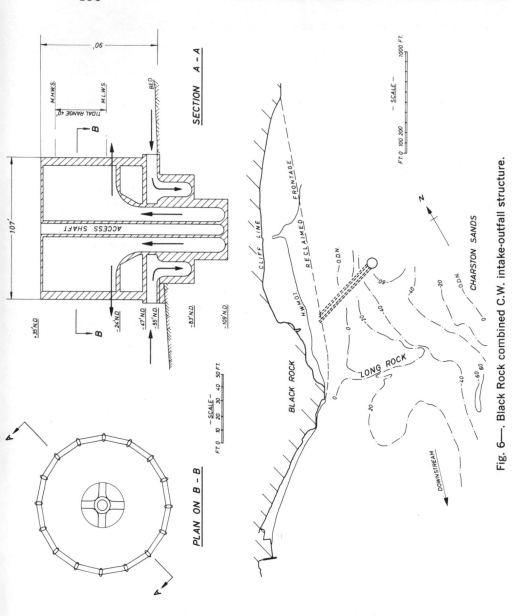

Fig. 6—. Black Rock combined C.W. intake-outfall structure.

Consequently, a natural scale model of the proposed arrangement was built. As shown in Figure 6, this was an annular structure with a screened intake all around its 111-foot diameter at an elevation − 47 feet N.D. to −55 feet N.D. The heated water rose up the annulus between a 34-foot-diameter cylinder and a 12-foot shaft to discharge over a horizontal apron at elevation −24 feet N.D. Consideration was also given to using a buoyant apron that would follow the tides with its top surface 1 foot below water level.

This model was operated to Froude scales, with equal density differences produced by using a unit scale of temperature rise. An area of 1200 feet by 700 feet was represented by the containing tank; but with no major cooling or exchange mechanism present in the model sump, there was a limit to the duration of any experiment because of the steady heat input into a restricted volume. This is a common difficulty with thermal models of a local zone. Results showed that a fixed apron gave less direct recirculation than a buoyant apron restricting the flow depth on it to one foot, and this was so with and without an ambient cross-flow.

Power Stations and Desalination Plant, Hong Kong

Two models were constructed for this investigation:
1. An undistorted model of the proposed intake and outfall for the proposed "C" station. The purpose of this was to solve minor problems connected with the hydraulic behavior of the spillway-type outfall, but principally to study the form, velocity, and dispersion in the jet from the submerged apron in the presence of cross-flows, and possible short-circuiting. This model was to a scale of 1/24, with a Froudian velocity scale of 1/4.9.
2. A small-scale, vertically exaggerated model including not only the "A" and "C" station cooling systems, but also the adjacent "B" station (see Figure 7) (Jaffrey and Gardiner, 1965). This covered a 7000-foot stretch of the harbor shore, extending 2200 feet offshore. The scales were 1/250 horizontal and 1/50 vertical, some 20 percent less vertical exaggeration than was strictly necessary for convective spread. This was not a full tidal model: it could be operated at selected steady conditions of level and cross-current.

The proving and adjustment of this type of model requires very detailed and complete prototype information. The adjustments for level and cross-flow are independent but are then correct for only that selected point in the tide cycle. This means that such a model can only be used to predict temperature distributions or recirculation at times in the cycle for which there is full field information about currents, with the possible addition of slack-water periods. Measurements of velocities, levels, ambient tempera-

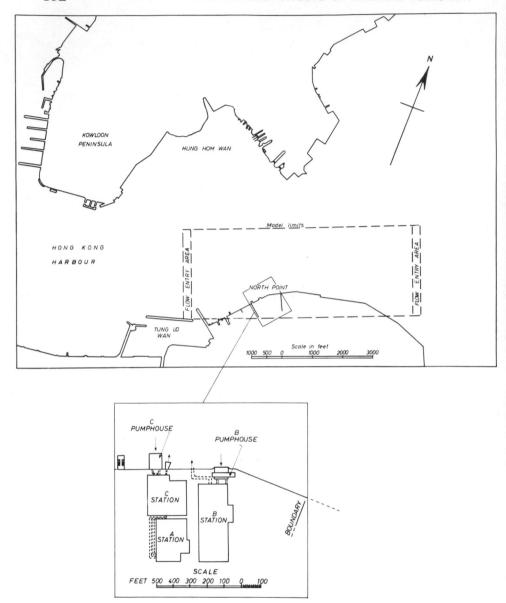

Fig. 7—. Hong Kong harbor power stations and model limits.

ture, and the pattern and temperature of cooling water emerging from the existing station should all be obtained simultaneously and, ideally, instantaneously. However, Hong Kong has a very complex tidal pattern. Practical limitations on the survey—the size of team and the amount of equipment

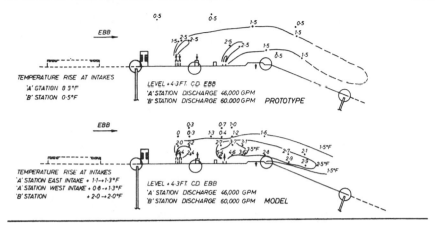

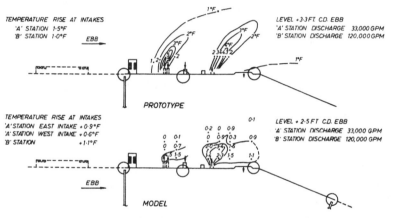

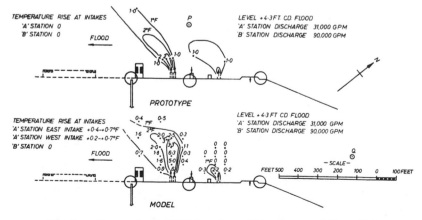

Fig. 8——. Comparison of field and model isotherms at Hong Kong.

available—thus meant that field results taken over a range of tides on different dates had to be grouped into similar conditions.

There was very little recirculation at the existing station and hence proving the model on the basis of temperature at the intake was a rather negative procedure. Also, the difficulties in getting adequate and representative thermal surveys in the field left considerable doubt when relating model to prototype about the validity of the comparison, bearing in mind that the prototype was responding to continually varying currents, whereas the model was running at a selected steady condition. Figure 8 (Jaffrey and Gardiner, 1965) illustrates the isotherms for the vertically exaggerated model and for the prototype at three of these "averaged" tidal states. The recirculation figures show that model predictions were conservative.

In the 1/24 model, shown in Figure 9, the cooling water was circulated direct from intake to outfall. The geometry of the outfall jet and turbulent mixing at its boundaries were expected to dominate the very local situation that could give rise to direct recirculation. Because an adequate supply of hot water to a model of this size was not practicable in the time available, it was decided to use a colorimetric technique to deduce recirculation in the absence of buoyancy forces. If buoyancy was significant in this local area, the model predictions would have been conservative, the intake being at a low level. Modifications were made to the original layout of the spillway and to the angle of the apron to reduce the observed short-circuiting. In all, seven modifications were tested, and the following table shows the inferred reduction in local recirculation with the selected design.

TABLE 1. Inferred Reduction in Recirculation: Outfall Structure, Hong Kong

Local recirculation results	Tide level and velocity			
	0 C.D. 0.5 knots	+ 3 ft C.D. 0.5 knots	+ 3 ft C.D. 0.75 knots	+ 6 ft C.D. 0.5 knots
14° apron slope, no step	4.8° F	2.8° F	3.3° F	2.1° F
14° apron modified as in Fig. 9.	1.5° F	1.3° F	2.3° F	1.8° F

Note: Outfall temperature 15° F. above ambient.

Thus improved, the outfall design was incorporated in the vertically exaggerated model. Observation confirmed that the horizontal spread in the jet-dispersion zone was too large and the vertical spread too small in the second model, and hence the outfall structure was provided with converging

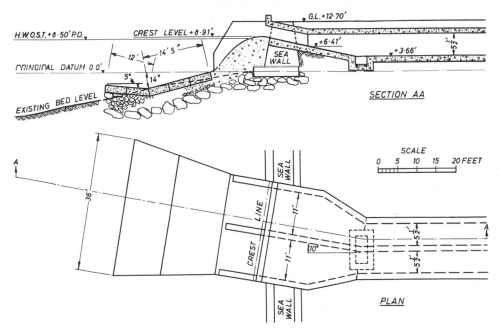

Fig. 9—. Outfall structure, Hong Kong. Simplified sketch of new "C" station
outfall as finally tested.

guide walls on the lower spillway and apron and the final section of the
apron was angled 5° below, rather than above, horizontal.

Thus, it was found possible by ad hoc experimentation to improve appreciably the modeling of the jet zone. Although the performance of the
two models still differed in some minor respects, there was now fairly good
agreement between them, as shown in the following tables 2, 3, and 4.

TABLE 2. Comparison of Velocities in Outfall Jet at Maximum Discharge.

		Apron toe	80 ft from sea wall			130 ft from wall		
Depth (ft)		2	2	4	6	2	4	6
Velocity in f.p.s.	1/24 Model	16.0	9.1	7.0	5.0	6.1	5.4	4.7
on axis of jet	1/250 Model	14.9	9.6	8.1	3.0	6.8	4.5	3.4
Velocity 10 ft	1/24 Model	—	1.0-2.0	—	—	3.0	—	—
either side of axis	1/250 Model	—	3.9	—	—	4.8	—	—

SOURCE: Jaffrey and Gardiner, 1965.

TABLE 3. Comparison of Maximum Velocities at Toe of Outfall Apron at Different Tidal Levels. No Crossflow. Max. Discharge.

	Tide level ft	0	1	2	3	4	5	6
Velocity in f.p.s. on	1/24 Model	16.0	14.5	13.6	12.5	11.2	10.0	8.6
jet axis at section 1	1/250 Model	14.9	—	12.5	—	11.1	—	9.0

SOURCE: Jaffrey and Gardiner, 1965.

TABLE 4. Comparison of Recirculation Figures

Level in ft relative to C.D.	Mean temperature rise at "C" intake in deg. F. above ambient	
	1/24 Model	1/250 1/50 Model
+ 6.0	1.8°	0.8°
+ 3.0	1.3°	1.25°
0	1.5°	1.35°

SOURCE: Jaffrey and Gardiner, 1965.
C.W. Discharge 140,000 g.p.m.
Crossflow 0.5k Flood. Measured 30 ft off center of "C" Pumphouse

In a model which does not cover the full tidal cycle, the assessment of the over-all efficiency of the cooling-water system is complicated because the results refer to selected times in the tidal period. A more useful comparison is possible if these separate results are combined to give the product of temperature rise and duration per tidal period or per day. Unfortunately, the tides in Hong Kong harbor, with their varying diurnal inequality, are far from ideal for this purpose, and to complicate matters further, flow off the North Point site is at times determined by the presence of a large inshore eddy. Figure 10 shows an example of how the tidal period was broken down into the flow zones covered approximately in the model.

Ringsend and Pigeon House Power Station, Dublin

In the previous examples of model studies, there was no necessity to pay special attention to salinity, but Dublin Harbor, which is used as a source of cooling water by the existing and proposed power stations on its flank, is strongly stratified, having a fairly small tidal volume in relation to the freshwater flow entering from the River Liffey.

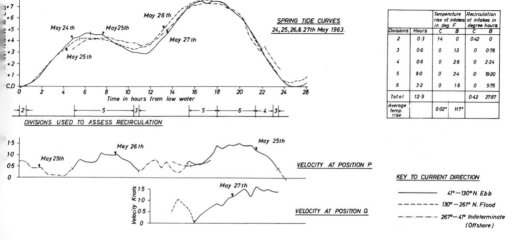

Fig. 10——. Estimates of recirculation on a spring tide, Hong Kong.

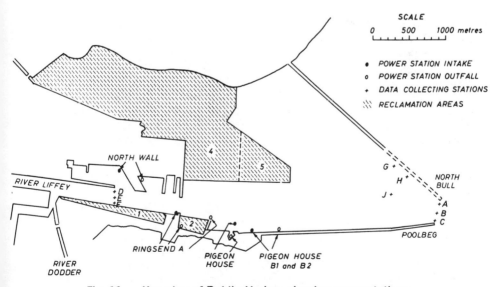

Fig. 11——. Key plan of Dublin Harbor, showing power stations.

The Electricity Supply Board's proposals involve trebling the existing Ringsend station to 270 megawatts, and replacing the present small station at Pigeon House by a capacity of 960 megawatts. The use of such a small body of water as Dublin Harbor for the discharge of a considerable heat load required stringent investigation to optimize the design and location of outfalls and intakes, taking into account major land reclamations within the harbor for port purposes. In addition to establishing the likely degree of recirculation, the extent of the "pools" of heated effluent and their temperature distributions had to be determined. The existing power station provided ideal information against which to prove the model, and the field survey carried out in 1963 covered:

1. Magnitude and directions of currents at surface, mid-depth, and near bed, throughout a spring and a neap tide at the points shown on Figure 11.

2. Salinities and temperatures under the same conditions at the same points.

3. Float tracks for some 6 hours around high water off the harbor entrance.

4. Thermal surveys during a neap tide with the existing power stations put on full-load, from 3 hours before L.W. to 1 hour after H.W. Temperatures were measured 1 foot, 4 feet, and 7 feet 6 inches below water surface. Eight such surveys were made during the period, at about 2-hour intervals.

The model was a rigid-bed one with scales of 1:300 and 1:50, using a pneumatic tide generator with modulation to simulate automatically the spring to neap cycle. The seaward end of the model was maintained at a salinity of 34 p.p. thousand, and the freshwater flows from the Liffey were added to produce the salinity gradients. The hot-water system was designed to give a 1/1 temperature scale, and because of the salinity variation with time, a scalar flow-through time was desirable in the heating circuit. Temperature measurement was by banks of thermistors (linearized over the range 10° C. to 27° C. with an accuracy of 0.05° C.) automatically held at pre-set depths below water surface: thermal readings at about 100 stations were used to deduce isotherms at H.W., ½-ebb, L.W. and ½-flood.

Adjustment of roughening produced good tidal reproduction and vertical salinity gradients, as shown in Figure 12. Simulation of the pronounced variation of salinity on a vertical is important in this case because the intake of denser water at a low level presents the possibility of the hottest layer of water lying below the surface.

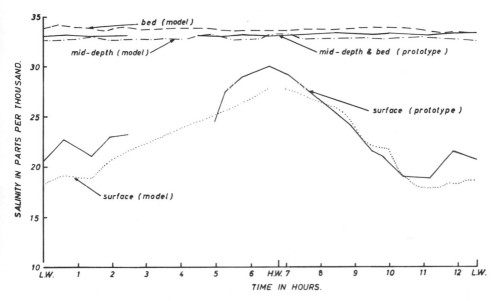

Fig. 12—. Prototype and model salinity profiles, Dublin.

The extensive thermal surveys provided an excellent basis of proving the model's performance. Model/prototype comparison with the existing stations at full load are shown in Figures 13 and 14 at two of the tidal states examined. This comparison, together with measurement of temperature at the intakes, clearly gave good confidence in the prediction of future conditions.

TABLE 5. Comparison of Intake Temperatures, Dublin.

	Ringsend A.	Pigeon House.
	Recirculation ° C hr/tide	Recirculation ° C hr/tide
Prototype	11.7	22.2
Model	11.0	23.4

Incidentally, over 140 separate tests were carried out, many being repeated more than once, in developing suitable outfalls and checking the effect of reclamations in the harbor.

Longannet Power Station on the Forth Estuary

Frazer, Barr, and Smith (1968) have described the model investigation

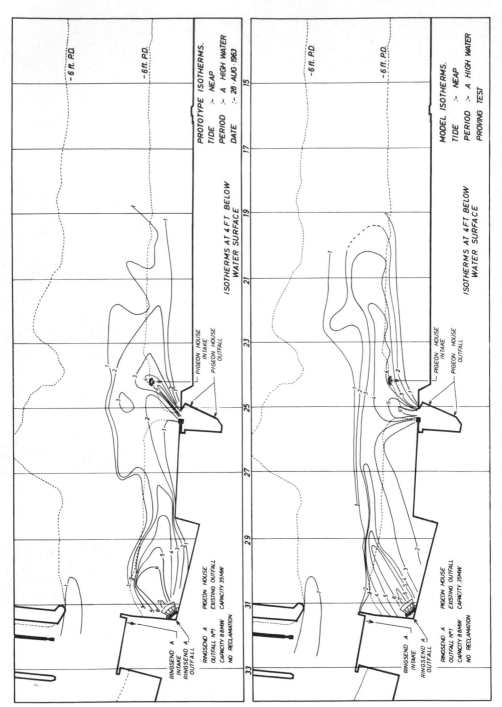

Fig. 13—. Comparison of model and prototype isotherms at high water. Temperature rise in °F. above background.

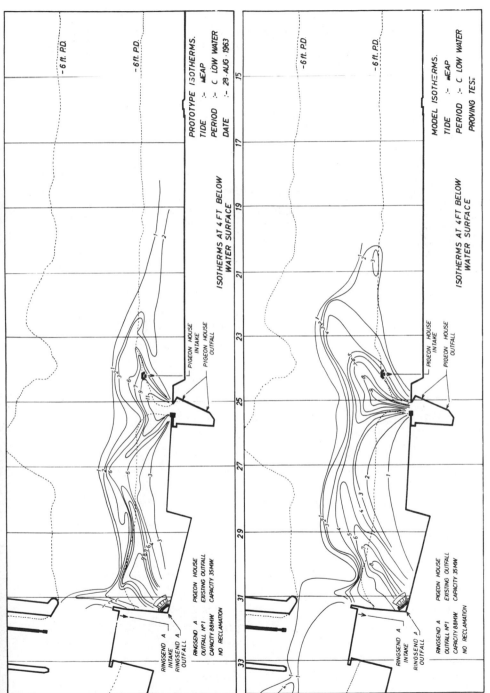

Fig. 14—. Comparison of model and prototype isotherms at low water. Temperature rise in °F. above background.

of heat dissipation from the 2400-mw station being built at Longannet in Scotland. This study by the University of Strathclyde involved a full tidal model of the estuary to scales of 1/400 horizontal and 1/60 vertical, (Figure 15). The published report of this study considers the simulation of surface profiles and salinity intrusion in tidal models and reviews the chosen scales in the light of the very recent results on lock-exchange flow obtained at Strathclyde in a flume 300 feet long, 5 feet wide with water depths up to 1.4 feet.

It has already been shown that simulation of convective spread involves a certain minimum vertical exaggeration, and the figures given for the Severn investigation demonstrate that, up to some calculable value of the extension of the front and selected vertical exaggeration, this will be accurately simulated, with increasing errors beyond that point. It was thought in the case of the Longannet station that the furthest extension relevant was to the far side of the estuary (the river authority was concerned about the possibility of a hot layer covering the full width). Taking 4000 feet extreme extension in 15 feet total depth of the analogous lock exchange phenomenon, the authors used their "congruency" diagram (Figure 5) to estimate the probable error in the model at that location; the figure they deduced of about 12 percent was considered acceptable. It is probable that the main risks from recirculation are at much shorter extensions of the convective front, for which in theory no error is involved in a Froudian model with unit scale of density difference.

Although the Forth estuary is well mixed, giving little vertical variation of salinity, the Longannet power station and its neighboring Kincardine station lie in a zone where there is a pronounced longitudinal salinity gradient. Thus, it was considered necessary to have a scalar flow-through time in the model heating system and to represent salinities in the model. Figure 16 shows the comparison of model and prototype observations (Frazer et al., 1968). The expected shift of the salinity curve upstream at low freshwater flows and downstream at flood freshwater flows was clearly demonstrated in the model. The longitudinal salinity pattern is dependent on the longitudinal turbulent dispersion coefficient: correct simulation of salinity thus indicates correct longitudinal dispersal, a factor which is also important in the behavior of effluents. Barr has shown that the longitudinal dispersion will be properly simulated if the dissipative energy is correctly scaled by adjustment of model roughness, a procedure that is normally carried out in order to get good tidal reproduction. The latter is thus also an indicator of the reproduction of dispersion.

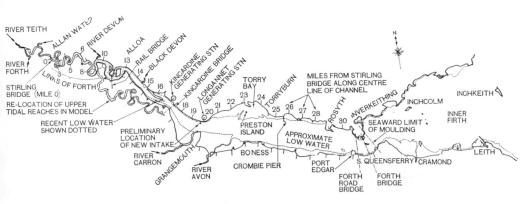

Fig. 15—. Tidal Forth. Key plan of tidal Forth,
showing relocation of upper reaches in model.
SOURCE: After Frazer **et al.**, 1968

Proposed 2,400-mw Station at Heysham, Lancashire

This review of some of the thermal models built and operated in the United Kingdom concludes with the investigation by the Hydraulics Research Station of the Heysham project. The intention is to withdraw 2,400 cusecs from within the harbor (see Figure 17) and to discharge it just outside the harbor beyond the tip of the west breakwater. The investigation is in its infancy, but the main tidal model to scales of 1/250 horizontal and 1/60 vertical was recently commissioned, and the proving stage, i.e., adjustment of tide and current reproduction, has just been completed.

Field work is to include a radioactive tracer study to determine the coefficients of turbulent dispersion. Corresponding experiments are to be made in the tidal model to check the reproduction of this phenomenon. This comparison should be of considerable interest.

Rather more progress has been made with a large-scale model without vertical exaggeration (scales 1/48) in which the buoyant plume and initial dispersion zones are being studied, at pre-set states of tide and current. About six variants of outfall are to be examined, the one illustrated in Figure 18 being a simple pair of 12-foot-diameter tunnels rising vertically to the sea bed. Figure 18 includes thermal results on a vertical plane across the direction of the ambient current, with only one outfall in action.

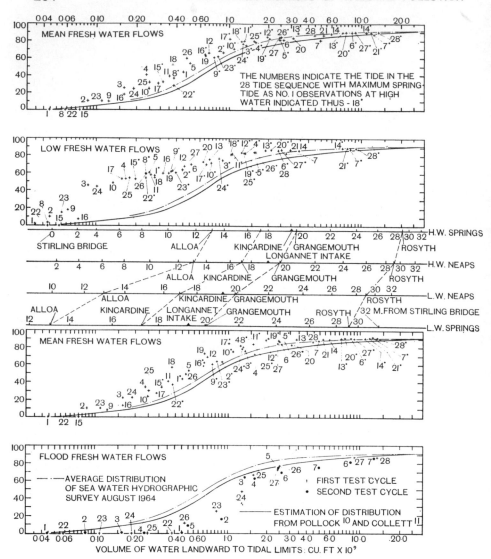

Fig. 16—. Model-prototype comparison of salinity intrusion, Forth, averaged through depth and shown as percentage seawater against landward volume.

SOURCE: After Frazer **et al.**, 1968

This model provides a good illustration of the very large heat requirement of large-scale thermal models. An insulated tank of 750-cubic-foot capacity is heated by 40-kw immersion heaters. Twelve hours of such input permit one hour's running of the model during the following day; so

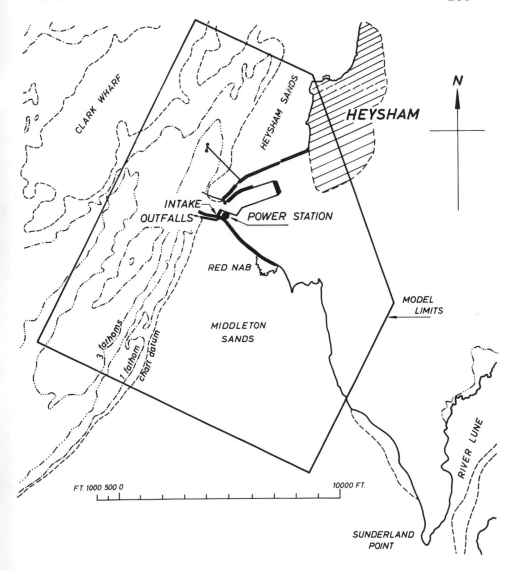

Fig. 17——. Heysham Harbor and power station site.

with this relatively short test duration, it is essential to plan the measurement program efficiently. Automatic data retrieval is most desirable.

Control and Instrumentation

A full account of the sophisticated instrumentation and automated con-

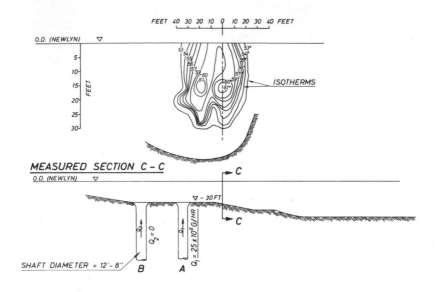

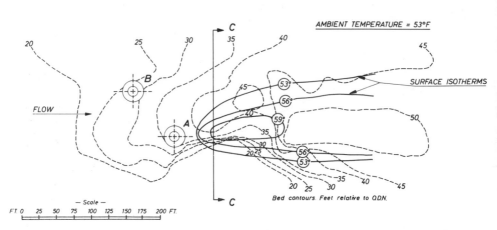

Fig. 18——. Outfall proposal for Heysham, with thermal survey. Heysham cooling-water investigation.

trol systems in use at Wallingford is impossible in a paper such as this. However, the underlying principles are:

1. To provide automatic generation of tides, coastal currents, fresh-water flows, and salinities so that ambient conditions are represented with the minimum of human intervention

2. To control accurately the thermal input from the model power sta-

tion, either on the basis of temperature rise, absolute temperature, or energy input, whichever is most appropriate in the circumstances

3. To monitor these imposed conditions via temperature, salinity, current, and level measurements

4. To measure and log the resultant conditions—particularly temperature, in the case of thermal models—with as dense a coverage in space and time as possible, consistent with instrument costs and the requirements of the problem

The following is merely an identification of the systems in use at Wallingford, including some special equipment for field surveys:

1. Automatic data-logging equipment is used at the station to record sequentially up to 100 variables of functions, such as water levels and temperatures. The inputs in the form of analogue voltages are converted from model to prototype scales and then into digital form for recording by strip printer as decimal characters or by paper-tape punch suitable for computer analysis. The time taken to scan 100 measurement points is about 30 seconds if printed, or 10 seconds on paper-tape.

2. Water-level transmitters are used in large numbers on tidal and flood investigation models, usually in conjunction with a data-logging system (see above). A sharp-pointed probe is caused to follow a changing water level by means of a servo-system. The instrument produces a voltage proportional to the water level which may be indicated and also recorded or digitized and printed, at a remote location. The range of the instrument is 8 inches (approximately 200 mm), the maximum rate of probe movement is about 0.2 in/s (5 mm/s) and the accuracy of following is better than ± 0.01 in. (0.25 mm). A later version of the instrument is being designed to have a range of 200 mm, an accuracy of 0.2 mm and a maximum rate of probe movement of about 100 mm/s.

3. The water-level transmitter/instrument carrier incorporates a servo-operated probe that follows a changing water level over a range of 1 foot (0.3 m), and carries additional transducers which it maintains at a constant depth below the water surface. For thermal investigations, eight temperature probes may be carried, at chosen points on plan.

4. The flow-recorder, which is servo-operated, continuously measures the head (h) over a Vee-notch weir; it then calculates the flow according to the equation $Q = 2.48 \ h^{2.48}$. The result is transmitted as an analogue voltage for remote recording.

5. The miniature-propeller current meter has been developed at the station for measuring flow velocities; it may be purchased from Kent Industrial

Instruments, Ltd., Gloucester Trading Estate, Gloucester, England. Measuring heads fitted with propeller rotors of 4-mm, 10-mm, and 20-mm diameter are suitable for measuring velocities in the range 20 to 1000 mm/s. A measuring head with a 10-mm rotor measures velocities in the range 50 to 5000 mm/s. The heads are used in conjunction with an electrical tachometer or, alternatively, a pulse-counting-and-timing unit.

6. The instantaneous ratemeter has been designed primarily to take full advantage of the response of the Miniature Propeller Current Meter to rapidly changing flows. The instrument continuously measures the reciprocal of the time interval between successive pulses in a chain of pulses. The measurements are obtained by digital methods but the readings are presented in analogue voltage form for recording. The range is 2 to 250 pulses per second with an accuracy of 0.1 percent of full scale or 1 percent of the reading.

7. Salinity-measuring equipment has been designed for use in hydraulic models. It measures the salinity of water within a very small volume (2-mm to 3-mm diameter) at the tip of a probe, and produces a voltage directly proportional to salinity (from 0 to 35 parts per thousand), which is independent of water temperature over the range 8° to 28° C.

8. The pneumatic tide-generator is economical in cost, flexible in use, and particularly suitable for very wide models. The tide box can be shaped to span (and to remove and deliver water across) the full width of a model.

A centrifugal fan continuously extracts air from an inverted box placed beside the model. The box is closed on all sides except for the side facing the model, where it terminates a short distance below the lowest low-water level; this leaves a gap through which water can flow between the box and the model. A controlled supply of air enters the box through a servo-operated valve which, by varying the pressure within the box, governs the amount of water drawn from or released into the model.

The pneumatic tide-generator may be used on models with salinity gradients, since the tide box can be compartmented so that water of different salinities, drawn from different depths in the model, may be stored separately.

9. The controlled-weir tide-generator is used whenever large discharges are required across narrow model widths or when two or more tide-generators are required on a single model.

The system consists essentially of a pump which supplies a continuous flow of water to the downstream (seaward) limit of a model at a rate slightly in excess of the maximum model-demand during the tidal cycle;

the surplus supply over the instantaneous demand is spilled over auto-matically controlled weirs to a sump, whence it is recirculated to the model.

When two or more generators are required on a model, one or more of them may be converted to discharge control by controlling the head over the weirs and also by supplying all the generators from a single pump through a constant-head tank. The cam output, representing the required time-varying head over the weirs, is compared with the actual head obtained by subtracting the weir-crest level from the water-surface level.

10. The littoral current-generator takes the form of axial flow pumps connected in parallel across the periphery of estuary or harbor models. This arrangement is used whenever it is necessary to reproduce currents that flow along the coast in some phase relationship with the varying tidal levels. A control system governs the speed and direction of the pumps so that they circulate water through the model in accordance with a pre-set program.

11. The curve-reading apparatus (electronic cam) provides a flexible form of input information to model control systems, such as tide-generators and current-generators, when the information is to be frequently modified during experiments. It consists essentially of a servo-controlled sensing head which aligns itself over a continuous line drawn in pencil or ink on a chart that is passed beneath the sensing head at a constant speed to match the time-scale of the model.

12. The temperature-control system is used on models designed to in-vestigate the recirculation of cooling-water from power stations. Water is pumped from the power station intake through a heater tube and past a temperature-sensing device; it then passes through a flow meter to the out-fall. Electrical power is fed to immersion heaters in the tube from thyristor circuits which vary the power according to the error between the required temperature and the measured temperature.

13. Thermal survey equipment is used to measure the temperature of thermistors towed from a boat. It prints the readings on a paper-strip chart, together with other data required to fix the time and location of the meas-urements. Over the range of 30° to 90° F., the accuracy is within $\pm$ 0.2° F. The equipment is portable and operates from a 24-d.c. supply.

14. Equipment for photographing a radar screen is used when float tracks are required in coastal waters. Small floats fitted with corner re-flectors can be detected by radar and their movements followed. The equipment photographs the radar screen at regular intervals so that the tracks of the floats can be plotted later. The photograph shows a group

of floats close to the survey boat; it also shows the distant coastline. A serial counter, date card, clock, and compass appear in the photograph.

A problem that is common to both field and laboratory temperature measurement is the acquisition of sufficient data over a short part of the tidal period to permit the preparation of sound instantaneous isotherm plots. In a well-controlled model, it is, of course, possible to get any density of coverage by sufficient repetition of test runs with the thermistors moved to new positions, but there is an obvious practical limit when one bears in mind that the investigation may have to cover many combinations of tide range, freshwater flows, stages of power station development and loading, adjacent heat sources, and types of outfall. In the field, cost imposes a very serious restriction to data collection by conventional techniques: one survey craft operating for twelve hours cannot collect adequate data for isotherm plots showing the spread of effluent in an area with strong tidal currents like the Severn.

But the United Kingdom's Central Electricity Generating Board have already made use of airborne infra-red techniques. This technique is sure to supersede ship-borne methods for obtaining surface isotherms: it can provide a close detail of surface temperatures almost instantaneously over the whole area affected by a power station, and semi-automatic methods exist for converting the scan into isothermal maps. Obviously, similar infrared scan techniques could be developed for use on a model scale, and first indications are that such a development would not be too difficult.

Summary and Conclusions

This report on hydraulic model techniques and the examples given of thermal models operated in the United Kingdom provided no more than an outline of what is a complicated field of study. The technique must be supported by adequate field surveys, and its limitations must be realized. Some of the imperfections of modeling are insuperable, but continued research on densimetric and turbulent processes must help improve the interpretation of scale-model results. Several of these fundamental studies are included in the bibliography, but this is admittedly incomplete.

The model-prototype comparisons that have been given show that reasonably good agreement is possible, although there remains the problem of the representation of initial mixing zones in vertically exaggerated models. The use of two models, one being a large-scale local one of the immediate locality of the outfall, is therefore often advisable. Expenditure on field surveys, not only so as to establish ambient conditions, but also to find out about dispersion from existing outfalls, may form an appreciable proportion of total investigation costs, but it is essential to spend this money if we are to have confidence in model findings. Infra-red techniques provide a promising addition to survey methods, in model as well as in prototype.

Despite the imperfections of the tool, physical modeling thus provides a very useful addition to other methods of study, an almost essential addition, in the view of some generating authorities where major power-station development is proposed in areas with complex boundaries and flow patterns.

REFERENCES

Abraham, G. 1965. "Horizontal Jets in Stagnant Fluid of Other Density." *ASCE, Proc., J. Hydraulics Division* 91(HY4):139–154.

Allen, J. 1952. *Scale Models in Hydraulic Engineering.* London: Longman Green and Co.

Anwar, H. O. "The Behaviour of an Effluent in a Calm Ambient Fluid of Greater Density." (In press.)

Barr, D. I. H. 1967. "Part III, Large-Scale Experiments." *La Houille Blanche* 22(6): 619–632.

Barr, D. I. H. 1963. "Densimetric Exchange Flow in Rectangular Channels. Part I: Definitions, Review, and Relevance to Model Design; Part II: Some Observations of the Structure of Lock Exchange Flow." *La Houille Blanche* 18(7):739–766.

Barr, D. I. H. 1963. "Model Simulation of Vertical Mixing in Stratified Flowing Water." *The Engineer* 215(Feb. 22):345–352.

Barr, D. I. H. 1963. "Model Simulation of Salinity Intrusion in Tidal Estuaries." *The Engineer* 216(Nov. 29):885–893.

Barr, D. I. H. 1958. "A Hydraulic Model Study of Heat Dissipation at Kincardine Power Station." *Instn. Civ. Engrs., Proc.* 10(July):305–320.

Bradley, J. N., and J. E. Warnock. 1948. "A Tidal Estuary Problem in Connection with the Design of the Antioch Steam Plant." 2nd meeting I.A.H.S.R.

Cederwall, K. 1968. "Hydraulics of Marine Waste-Water Disposal." *Chalmers Institute of Technology,* Hydraulics Division, Goteburg, Sweden, Report No. 42.

Dedow, H. R. A. 1965. "The Control of Hydraulic Models." *The Engineer* 219 (Feb.):301–304.

Ellison, T. H., and J. S. Turner. 1959. "Turbulent Entrainment in Stratified Flow." *Journal Fluid Mechanics* 6(Oct.):423–448.

Fan, Loh-Nien, and N. H. Brooks. 1966. "Discussion of 'Horizontal Jets in Stagnant Fluid of Other Density' by G. Abraham." *ASCE, Proc., J. Hydraulics Division* 92(HY2):423–429.

Favre, J. 1961. "Similitude Conditions in Thermal Power Station Cooling-Water Circuit Studies." *La Houille Blanche,* No. Special A. (In French.)

Frankel, R. J., and J. D. Cumming. 1965. "Turbulent Mixing Phenomena of Ocean Outfalls." *ASCE, Proc., J. Sanitary Eng. Div.* 91(SA2):33–59.

Frazer, W., D. I. H. Barr, and A. A. Smith. 1968. "A Hydraulic Model Study of Heat Dissipation at Longannet Power Station." *Proc., Institution Civil Engineers,* 39(Jan.):23–45.

Garrison, J. M., and R. A. Elder. 1965. "A Verified Rational Approach to the Prediction of Open-Channel Water Temperatures." *Proc., Eleventh Congress, I.A.H.R.,* Leningrad.

Harleman, D. R. F., and R. A. Elder. 1965. "Withdrawal from Two-Layer Stratified Flows." *ASCE, Proc., J. Hydraulics Division* 91(HY4):43–58.

Harleman, D. R. F., and K. D. Stolzenbach. 1967. "A Model Study of Thermal Stratification Produced by Condenser Cooling-Water Discharge." *M.I.T. Hydrodynamics Laboratory Report No. 107.*

Jaffrey, L. J., and I. M. Gardiner. 1965. "Studies in Field and Model of Cooling-Water Circulation in Hong Kong Harbour." *Proc., Eleventh Congress, I.A.H.R.,* Leningrad.

Jaffrey, L. J., D. M. Williams, P. Ackers, and W. A. Price. "Discussion of 'A Hydraulic Model Study of Heat Dissipation at Longannet Power Station, by W. Frazer, et al.' " (In press.)

Lean, G. H., and A. F. Whillock, 1965. "The Behaviour of a Warm-Water Layer Flowing over Still Water." *Proc., Eleventh Congress, I.A.H.R.,* Leningrad.

O'Brien, M. P., and J. Cherno. 1934. "Model Law for Motion of Salt Water Through Fresh." *Trans., ASCE* 99:576–609.

Price, W. A., and M. P. Kendrick. 1962. "Density Currents in Estuary Models." *La Houille Blanche* 5(10).

THE author's outline of modeling principles is a good, concise presentation of the principles governing the simulation of stratified flow. As the author indicates, this is an outline and the necessary brevity prevents amplification of these principles. However, a few comments are thought necessary to emphasize some assumptions that have been made.

In the author's discussion of roughness to give correct head loss, it is stated that

$$\lambda_c = \lambda_v \, \lambda_x^{1/2} \, \lambda_y^{-1} \qquad \text{(Equation 5)}$$

In general, for a distorted model, $\lambda_x < \lambda_y$. Equation 5, therefore, shows that $\lambda_c < 1$ for a Froudian model for which $\lambda_v = \lambda_y^{1/2}$. Implicit in the derivation of Equation 5 is the assumption of an infinitely wide channel. In many, perhaps most, instances this is a valid assumption; however, it should be pointed out that, if the channel is infinitely deep, the ratio of Chezy coefficients becomes

$$\lambda_c = \lambda_v \, \lambda_y^{-1/2} \qquad \text{(Equation D1)}$$

For a Froudian model, Equation D1 shows that $\lambda_c = 1$. Depending upon the channel geometry, it is possible to have a value of λ_c between that given by Equation 5 and unity as shown by Equation D1.

The author also demonstrates that a distorted model is necessary to simulate heat exchange with the environment and shows that the vertical exaggeration is equal to the square root of the vertical scale ratio. The author mentions, and it is worth re-emphasizing, that the same climatic conditions must exist in the model and prototype. That this is so can be demonstrated by writing the expression for the longitudinal temperature gradient for a thermally homogeneous stream:

213

$$\frac{dT}{dx} = \frac{- K(T - T_e)}{\rho chu} \qquad \text{(Equation D2)}$$

where T is the temperature at some distance x, T_e is the equilibrium temperature, ρ is the density of water, c is the specific heat of water, h is the depth, u is the velocity, and K is the environmental heat-exchange coefficient. In terms of model-prototype ratios, Equation D2 can be expressed as

$$\lambda_T = \lambda_K \, \lambda_T \, \lambda_e^{-1} \, \lambda_c^{-1} \, \lambda_h^{-1} \, \lambda_u^{-1} \, \lambda_x \qquad \text{(Equation D3)}$$

Assuming a Froudian model, using water as the test media, Equation D3 reduces to

$$\lambda_K = \lambda_y^{3/2} \, \lambda_x^{-1} \, \lambda_\rho \qquad \text{(Equation D4)}$$

If the model is operated at prototype temperatures, then $\lambda_\rho = 1$; and if the climate in the model and prototype is the same, then $\lambda_K = 1$ and

$$\lambda_x = \lambda_y^{3/2} \qquad \text{(Equation D5)}$$

which agrees with the author's result. In general, however, the climate in the model will not be the same as the prototype, unless a special, environmentally controlled laboratory is available. For a model operated in a laboratory without benefit of environment control, the humidity is likely to be greater and the wind movement less than in the prototype. Both of these factors tend to reduce the evaporative heat loss in the model. One must also consider that, if the model is operated at the same temperature as the prototype, radiation emitted from the water surface in the model is the same as in the prototype; however, unless the model is out of doors and in the same geographical location as the prototype, incoming short- and long-wave radiation is not the same for model and prototype. In addition, the effects of short-wave radiation may be quite different in model and prototype. For lakes in the TVA area, short-wave radiation penetrates to a depth of approximately 10 feet (Wunderlich and Elder, 1968), which, in the prototype, may represent only a fraction of the total depth; however, this is likely to be more than the entire depth of the model. The amount of long-wave radiation received by an indoor model will depend upon the surrounding material and will vary from one laboratory to another. Without evaluating all of the terms in the heat-budget equation, it is difficult to make a general statement that a model will dissipate heat at a rate greater or less than the prototype.

It would seem that a rather simple experiment could be conducted simultaneously with the operation of a thermal model to determine the environmental heat-exchange coefficient applicable to the model. Comparisons to the prototype could then be made and, if the results warrant, modifications could be made either to the model scale or to the model results.

Mr. Ackers makes a statement that is well worth repeating: "Despite the imperfection of the tool, a physical demonstration of recirculation paths in the complex situation of estuary flow is of undoubted value, for only in the simplest circumstances can computations produce convincing evidence about likely thermal pollution." Analytic techniques have not developed to the point where anything but the simplest cases can be adequately analyzed. Existing analytic techniques serve to provide a basis for a preliminary estimate of the requirements for a condenser-water intake and outlet system; but, to insure a proper design, model studies in most cases will be necessary.

To illustrate, a brief description of a model study of the condenser-water discharge canal for the Cumberland Steam Plant that is currently being conducted at the Tennessee Valley Authority's Engineering Laboratory will be presented. The Cumberland Steam Plant is a conventional fossil-fuel plant located on the Cumberland River near Nashville. This plant will have two generation units of 1300 megawatts each. The condensers for this plant were designed for a 12° F. temperature rise and as such will require a total flow of approximately 4200 cfs. The plant site is located in the backwater of Barkley Dam, a multipurpose project located approximately 73 miles downstream. The depth of Lake Barkley at the plant site is approximately 45 feet. Cheatham Dam, also a multipurpose project, is located approximately 45 miles upstream. The hydroelectric facilities at Cheatham and Barkley are operated for electrical peaking purposes, thereby creating unsteady flows at the plant site. The Cumberland River is well regulated by upstream dams; however, because of the size of the Cumberland River watershed and its history of having two consecutive dry years, a guaranteed minimum flow of only 5000 cfs is available. During these periods of low flow, the steam plant will divert 84 percent of the flow.

A preliminary analysis of the site characteristics indicated that a skimmer wall, combined with a non-mixing discharge canal, would be the best

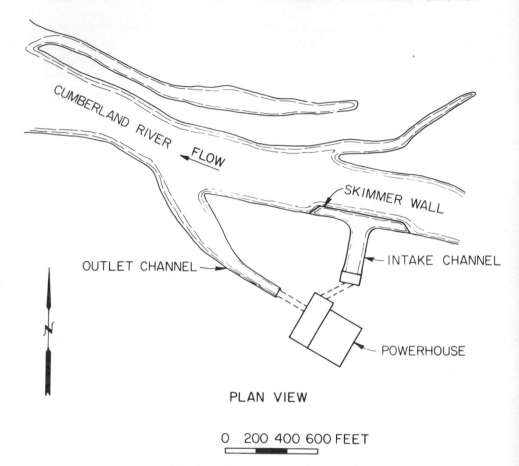

PLAN VIEW

0 200 400 600 FEET

Fig. 1—. Cumberland Steam Plant.

system for preventing recirculation and maintaining acceptable thermal conditions in the receiving water. By designing a discharge canal with a densimetric Froude number less than one, the warm water will be discharged as a thin layer over the cooler receiving water. The thickness of the warm-water layer can be controlled by the geometry of the canal.

The outlet channel and skimmer wall represent an integral system for preventing recirculation. Both must perform in the manner for which they were designed if the system is to be effective. Based upon previous laboratory investigations (Harleman and Elder, 1965), we were confident that the skimmer wall would perform as designed; however, there were many questions as to the geometry of the discharge canal. As shown in Figure 1, the discharge canal, as originally conceived, consisted of a rapidly divergent section from the condenser-discharge tunnels to a trapezoidal open channel. The channel then ran straight for approximately 500 feet, to where it began a bend and expansion. The channel was then straight for a distance of approximately 600 feet, to where it entered the Cumberland River. This configuration was adopted by the designers primarily because of economics dictated by the topography of the site. On review of this configuration, we became concerned about the possibility of flow separation at the bend and the possibility that undesirable mixing of the warm and cold water would occur. If excessive mixing should occur, the canal would not perform as designed. A review of the literature revealed a paucity of information of either an analytical or empirical nature which would be of assistance in the solution of this complex problem; therefore, it was recommended that a model be constructed to investigate the performance of the canal. Since the primary purpose of the model is to investigate flow patterns and mixing, an undistorted densimetric Froude model is necessary. It is also necessary that the friction factors in the model and prototype be the same in order to simulate properly the head loss in the channel.

To demonstrate that a distorted model will not properly simulate mixing, consider a prototype in which hot water is discharged through a rectangular slot into a quiescent body of colder water. Because of the density difference between the jet and the receiving water, buoyancy forces deflect the jet upwards, the trajectory being determined by the density difference and the initial velocity of the jet. Assume that two models are built to investigate this prototype. One model is undistorted, with a length-scale factor of λ_{x_1}; the other model is distorted, with a vertical-scale ratio of $\lambda_{y_2} = \lambda_{x_1} = \lambda_{y_1}$ and a horizontal-scale ratio of λ_{x_2}. Both models are operated at prototype temperatures. Because the velocity ratio is determined by the square root of the vertical-scale ratios, both models will have the same initial jet velocities; therefore, since the density differences are the same, the trajectory will be the same in both models, and the location at which the jet reaches the surface will be the same in both models. But, when converting from model to prototype horizontal distances, the undistorted model gives a length of

$$L_p = L_M \left(\frac{1}{\lambda_{x_1}} \right) \qquad \text{(Equation D6)}$$

whereas, the distorted model yields

$$L_p = L_M \left(\frac{1}{\lambda_{x_2}} \right) \neq L_M \left(\frac{1}{\lambda_{x_1}} \right) \qquad \text{(Equation D7)}$$

To show that a unity-friction factor is necessary to simulate head loss, write the Darcy-Weisbach equation

$$h_l = f \frac{l}{d} \frac{V^2}{2g} \qquad \text{(Equation D8)}$$

To satisfy the modeling criteria

$$\lambda_{hl} = \lambda_f \lambda_l \lambda_d^{-1} \lambda_v^2 \qquad \text{(Equation D9)}$$

For an undistorted densimetric Froude model

$$\lambda_{hl} = \lambda_l = \lambda_d = \lambda_x \qquad \text{(Equation D10)}$$

$$\text{and} \quad \lambda_v^2 = \lambda_x \qquad \text{(Equation D11)}$$

$$\text{therefore} \quad \lambda_x = \lambda_f \lambda_x \qquad \text{(Equation D12)}$$

$$\text{or} \quad \lambda_f = 1 \qquad \text{(Equation D13)}$$

The scale of the model was determined by the requirements that both the friction factor and densimetric Froude number be equal in model and prototype. The prototype absolute roughness, k_p, was estimated, based upon a knowledge of the type of material and the method of excavation. Knowing the hydraulic radius of the prototype, the dimensionless parameter $4R/k_p$ was computed, and the friction factor, f, determined for the prototype Reynolds number, IR_p. Figure 2 shows that the model Reynolds number, IR_M, required to give a friction factor equal to the prototype value decreases with a decrease in the roughness of the model surface. To obtain a convenient model-size, the model was constructed from hydraulically smooth plexiglass.

Since

$$IR = \frac{VL}{\nu} \qquad \text{(Equation D14)}$$

$$\lambda_{IR} = \lambda_V \lambda_L \qquad \text{(Equation D15)}$$

where ν is kinematic viscosity, which is the same in model and prototype, V is a characteristic velocity, and L is a characteristic length. To satisfy the densimetric Froudian requirement

$$\lambda_V = \lambda_L^{1/2} \qquad \text{(Equation D16)}$$

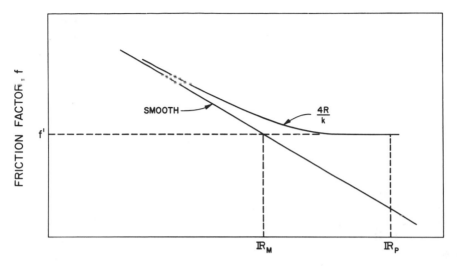

FRICTION FACTOR, f

f'

SMOOTH

$\dfrac{4R}{k}$

$\mathbb{R}_M$ $\mathbb{R}_P$

REYNOLDS NUMBER, $\mathbb{R}$

Fig. 2—. Resistance Diagram.

Equation D15 reduces to

$$\lambda_L = \lambda_{I\!R}^{2/3} \qquad \text{(Equation D17)}$$

Density in the model can be controlled by using either a saline solution to represent the denser, colder water, with fresh water to represent the warm water, or by actually using warm and cool water. It was decided that the use of warm water would be more satisfactory. Several reasons were responsible for the choice of a thermal model. One of the most important reasons is that, with a saline-induced stratification model, the interface is sharper and more stabile than the interface in a hot-and-cold-water model (Harleman and Stolzenbach, 1967). A second reason is the relative ease and cost of measuring temperature, as compared to conductivity. Additionally, the problems of corrosion which are encountered with a saline solution are avoided with the thermal model.

The model is shown in Figure 3, together with the hot-water supply system. As indicated, water is pumped from a warm-water collection box by a centrifugal pump. The flow is split on the discharge side of the pump, and a portion of the flow directed through the heat exchanger; the remainder goes directly to a mixing valve. The hot-and-cold-water supply lines rejoin at the mixing valve, which regulates the temperature of the condenser discharge. The desired temperature bias between the river and

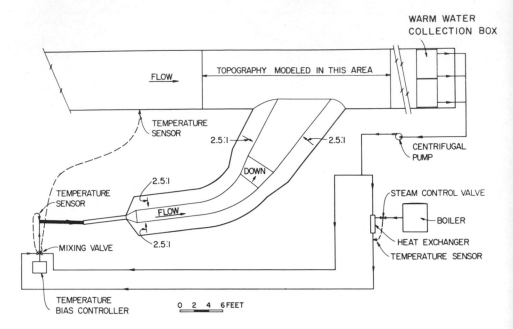

Fig. 3—. 1:55 model of Cumberland Steam Plant condenser-water discharge channel.

condenser discharge is set at the control box, and the valve is regulated automatically by the temperature sensors located in the main channel and in the pipe manifold. To prevent excessive hunting by the mixing valve, a second control valve was installed in the steam line. This valve regulates the temperature of the water leaving the heat exchanger. With this control system, the temperature can be maintained within one degree of the desired temperature.

To prevent warming of the sump temperature, which the author indicates was a problem in the Black Rock outfall/intake model, a warm-water collection box was designed to skim off and recirculate the warm surface layer and allow the cooler lower layer to return to the sump. This system has proven effective in preventing heat-up of the sump water.

Data from this model consists of photographs of dye releases and temperature measurements at selected locations. The present temperature measuring equipment consists of thermistors connected to a stepping switch which can be programmed to sample any combination of points, either automatically or manually, at either pre-set time intervals or on command.

These data are printed on a standard teletype and simultaneously punched on paper tape in a computer-compatible code.

Operation of the model immediately justified our concern. Three problem areas were identified: (1) the rapidly divergent transition at the up stream end of the discharge canal did not distribute the flow properly, (2) flow-separation occurred at the bend, and (3) undesirable mixing occurred near the mouth of the canal. To determine the effect of the transition on the problems observed at the bend and near the mouth, a baffle was installed at the transition. The baffle was effective in straightening the flow, and its effects were noticeable in a reduced-flow separation and mixing; however, it did not eliminate these undesirable conditions.

Considerable work remains to be done to determine the best configuration for the discharge canal.

REFERENCES

Harleman, D. R. F., and R. A. Elder. 1965. "Withdrawal from Two-Layer Stratified Flows." *ASCE, Proc., Hydraulics Division* 91(HY4):43–58.

Harleman, D. R. F., and K. D. Stolzenbach. 1967. "A Model Study of Thermal Stratification Produced by Condenser Cooling-Water Discharge." M.I.T. Hydrodynamics Laboratory Report No. 107.

Wunderlich, W. O., and R. A. Elder. 1968. "Evaluation of Fontana Reservoir Field Measurements." Paper read at ASCE Specialty Conference on Current Research in the Effect of Reservoirs on Water Quality, January 1968, at Portland, Oregon.

DISCUSSION FROM THE FLOOR

Donald Harleman: I think one of the important things that Mr. Ackers has shown us is the comparison between the model and prototype data. This is much more difficult to obtain in this country, since we have done a smaller number of these studies. I do not know of very many instances where we have comparable sets of data. We should encourage more follow-up studies of completed plants which have previously been modeled, so that we begin to collect more confidence in dealing with these problems experimentally. I would like to inject a word of caution, especially in dealing with tidal-estuary models where salinity intrusion may be a factor. The word of caution is necessary, because one is able to reproduce salinity intrusions by a cut-and-dried process of making the model do what you observe in the prototype (by roughness elements or strips to promote mixing). These elements are designed to promote mixing from the bottom up, to bring the salt up into the upper layer. In the same model, if you discharge heated water, the problem would seem to be one of mixing down, since the heated water tends to rise. It is not at all obvious that by reproducing salinity distribution you can expect to have proper thermal mixing. This is certainly one of the difficulties of dealing with the local mixing zone.

R. E. Nakatani: I wish to ask Mr. Ackers for comments about future research and development activities with these physical models. Do you have fishery biologists working closely as a team with the hydraulics engineers on these models? Recently, Mr. M. Perry, a fisheries officer of the Yorkshire Ouse and Hull River Authority, in your country, requested information on our experiments related to behavior of adult steelhead and Chinook salmon near Hanford's reactor plumes. He was concerned about the effect of a future coal-fired power station to be constructed near an estuary of a river with a run of Atlantic salmon.

He was especially interested in temperature patterns which may cause a thermal block of the salmon. Would you please comment?

Peter Ackers: The only way I can comment is to describe very briefly what the administrative set-up is in Britain, which is rather different from yours. Most of our work in the thermal-pollution field comes to us from the electricity-generating board. We don't have private power companies, over there. The generating board has to satisfy the river authorities as to the degree of thermal pollution that might arise from a proposed development. These river authorities are, of course, based on the catchment areas, and there is an avenue of appeal to the government's level, to the Ministry of Agriculture of Fisheries and Food. These river authorities have their fishery officers, and this man, Mr. Perry, presumably from the Yorkshire River Authority, is such an expert. If the Central Electric Generating Board is seeking to develop a power station in Yorkshire, they might well come to us to have the hydraulic model-study done. We will make some sort of estimate on their behalf as to where the hot water will be, and how hot it will be, and then it's up to the Central Electric Generating Board to convince the river authority that this is acceptable. Clearly, if the generating board failed to convince the river authority, then this information is fed back to us, and we are told the scheme that was put forward is unacceptable to the fishery people of the river authority, possibly because there is a layer of hot water across the full width of a river at a particular site. They would now discuss ways of overcoming this, which would mean some sort of a second-stage study. The direct answer to your question is no, we do not have fishery experts on our staff. This is not part of our station's work, but obviously the river authority fisheries people get involved before approvals for power-station developments are given.

Carlos Fetterolf: I would like to ask Dr. Edinger what the size of the cooling pond was which he illustrated for us, and what the heat input to that pond was. I would also like to ask what percent of the heat loss was to the atmosphere, and if cooling towers had been used to dissipate an equal amount of heat, how much more moisture would have been added to the atmosphere by the cooling tower than would be added by the evaporation from the pond or other cooling devices. What I am trying to get at is the relative losses of water to the atmosphere by various cooling methods.

John E. Edinger: The particular case illustrated on that first slide was a pond with an area of about 430 acres, related to a plant size of discharging 35.6 billion BTUs a day. For the percentage of heat loss, all you would have to do is take that in proportion to your temperature rise leaving the pond. The only place heat could be lost is to the atmosphere. On relative magnitudes of the heat-loss terms, upwards of 80 to 85 percent of the heat loss could be by evaporation and the remaining 15 percent by radiation and conduction. This varies depending on the general temperature range; for example, if the body is between 80 and 90 or between 90 and 100 or between 70 and 80.

William Spencer Davis: Biologists are concerned with the absolute value and duration of extreme temperatures, not average temperatures. Is it possible to virtually eliminate heat transfer from water to the atmosphere for a matter of days or weeks, during hot seasons of the year, based on the atmospheric conditions—especially equilibrium temperatures—and what are the space and time factors involved in this?

Edinger: On a weekly trend, say in the spring months, yes; but within this period, there would still be diurnal cooling.

Davis: When the temperature of the air is quite high, much higher than the temperature of the water, and remains high, and the water temperature, with the input of waste-heat, never quite gets up to that point, are we going to lose heat to the atmosphere, or not?

Edinger: We are always going to be losing heat to the atmosphere, in any case, when the water temperature is greater than the equilibrium temperature.

Davis: If the water temperature is less than the equilibrium temperature?

Edinger: Then we are gaining heat from the atmosphere to the water temperature.

Davis: During the not-dry season of the year, is it possible for the temperature of the water with the waste heat input to remain below the atmospheric temperature?

Edinger: Sure, because it's below the water temperature 12 hours a day. However, it's the short-wave radiation which influences the diurnal cycle in the equilibrium-temperature relations and governs its sharp diurnal fluctuation. The air-temperature cycle on it contributes a few degrees to the equilibrium-temperature variations, but it's nighttime cooling where most of the heat is lost.

THE HORIZONTAL TRAVELING SCREEN

THE passage of young fish through power plants and into unscreened irrigation diversions results in heavy, undesirable mortality. To prevent these losses, the U.S. Bureau of Commercial Fisheries has been searching for an efficient, relatively low-cost plan for deflecting the fish away from these potentially dangerous areas.

Figure 1 illustrates the direction of our work. We desire to deflect fish

EFFICIENTLY COLLECTING
ALL MIGRANTS
LARGE FLOW VOLUMES—
MINIMAL COST

Fig. 1—. This states the purpose of our work, or our mission.

away from the source of danger and to do so without causing injury to the fish; to collect all fish, even the smallest; to collect fish from large volumes of flow, such as rivers; and to accomplish this at minimal cost.

Figure 2 shows the Leaburg Canal, a power-plant diversion of the McKenzie River, Oregon. This canal, having a flow of 2,400 second feet, measures 70 feet wide at the surface, 40 feet wide at the bottom and 17 feet in depth.

Figure 3 illustrates the design of a standard water screen, the industrial type. The cost of screening the relatively small Leaburg Canal shown in

Fig. 2—. The Leaburg Canal, a diversion of the McKenzie River.

Figure 2 with industrial water screens would be $500 per second foot, or a total of $1,250,000.

Figure 4 shows an industrial water screen in cross section and illustrates the bulkiness of the facility. We believe this screen is too heavy for our needs, does not travel with sufficient speed, and travels in the wrong direction.

Figure 5 shows the upstream side of a fish-deflecting facility used on the West Coast and referred to as louvers. Since there are no traveling parts in this facility, the maintenance is minimal.

Figure 6 shows the reverse side of the 40-foot-high louver structure. As you can see, the design is quite involved. This structure screens a power plant intake flow of 6,000 second feet. There are two such facilities. The louver structure is located on the Cowlitz River in the state of Washington.

A major disadvantage of the louver system is the extreme variance of fish-deflection efficiency. With certain sizes and species, losses can be as great as 70 to 80 percent; in contrast—again, as a factor of size, species, and design of the louver system—deflection efficiencies of 97 to 99 percent can be considered normal.

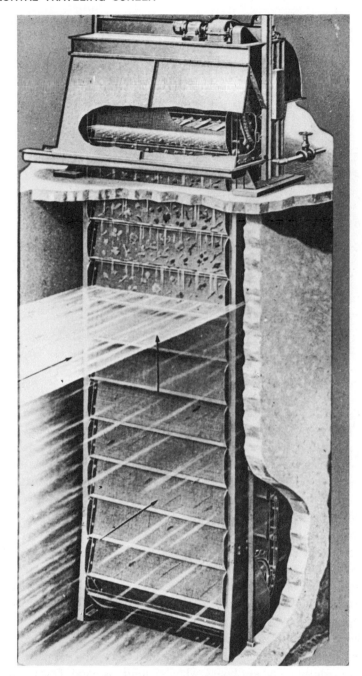

Fig. 3—. This illustrates the design of a standard industrial water screen.

Fig. 4——. Sectional view through the bottom end of an industrial water screen.

Over the years, we have tested many fish-guiding devices, including sound. The advantage of sound, were it to be successful, would be in the limited amount of hardware required. Unfortunately, up to now, no success whatsoever has been achieved in deflecting young fish. Figure 7 shows a test stand in which a cooperative fish-sound study was conducted by the Bureau of Commercial Fisheries and the Boeing Company of Seattle, Washington.

Figure 8 pictures a prototype experimental electrical fish screen located on an irrigation diversion in the state of Washington. Previous studies on electrical guiding conducted by the Canadians resulted in their complete abandonment of the plan. The variables of electrical guiding are many and complicated. The investigations have been dropped by both governments for the time being.

Shown in Figure 9 is a curtain of air bubbles used in an effort to deflect fish. The bubbles floating up from the canal floor make a fairly effective fish barrier during daylight hours, when visibility is good. During the hours of darkness, however, with loss of visual acuity, fish penetrated the bar-

Fig. 5——. Downstream view of the Mayfield louvers and fish bypass.

rier without apparently realizing it. The resultant deflection efficiencies were unacceptable. Lighting the area was ineffective, since visual adjustment time for fish passing from dark to light is a matter of many minutes, in contrast to seconds for man. Barriers formed of water jets were no more successful.

On the extreme left side of Figure 10, we illustrate the manner in which fish screens were installed years ago: normal to flow. Downstream migrants on their way to the ocean, on reaching this facility, were required to stop and search in an effort to find the bypasses provided. This caused delay and an undue expenditure of effort.

In the center of this illustration, we show the next advanced step— the placement of a fish-deflecting structure on an angle of 30 degrees to the direction of flow. Fish passing downstream, on approaching the structure, position themselves head-first into the flow. This apparently provides them with increased control over their movements, making it easier to avoid obstacles as they drift downstream.

When the structure is placed on an angle to flow, fish moving downstream and into the bypass, as shown here, do so without hesitation. On the extreme right, we show our latest plan for deflecting fish: the traveling

Fig. 6——. View of the Mayfield louvers from the downstream side.

Fig. 7——. Test equipment used in the deflection of fish by means of sound.

Fig. 8——. A prototype experimental fish screen.

Fig. 9——. Air bubbles used in fish-deflection studies.

Fig. 10——. Illustrations from left to right show
advances achieved in the placement and design
of fish guidance systems.

screen. Travel of the screen is across the full width of the canal. At this point, it makes a 180-degree turn traveling back upstream. The netting hangs full depth.

Figure 11 is an artist's concept of the traveling screen. It illustrates the screen's simplicity of design and the reason why we believe construction costs will be minimal. The only portions of the structure in the water are the stiff legs, spaced approximately 10 feet apart, and panels of wire-cloth screen or nylon netting. There are no drive systems, cables, or connectors on the canal floor to cause difficulty. The bottom of each screen panel comes to within 2 inches of the canal floor. This interval of space is sealed off by means of nylon brushes attached to the bottom of each panel to provide a seal against fish loss. To prevent the stiff legs from deflecting downstream, as a factor of water pressure, a cantilever support system has

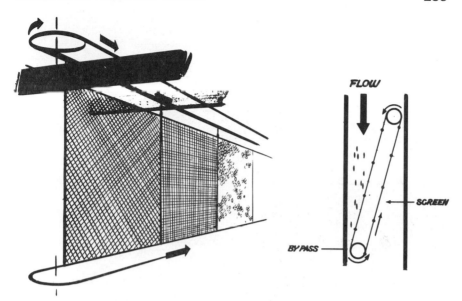

Fig. 11——. On the left, an artist's concept of the design of the traveling screen.

On the right, a plan view of the traveling-screen concept.

been installed at a point up and out of the water. Here we show the track, the carriages which are connected to the stiff legs, and the drive system. This is a simple but effective screen.

When a traveling screen moves across a canal on an angle to flow and returns back upstream, the fish move away from the screen and into the bypass. The traveling screen allows for the use of approach velocities up to 4 to 5 feet per second, dependent on fish size. With screens placed normal (90°) to flow, approach velocities have generally been limited to 0.5 to 1.5 feet per second, again dependent on species and size of fish. By placing the screen on a 20- to 30-degree angle to flow, velocities of a much greater magnitude can be used. For example, with a screen placed on a 12° angle and migrant fish capable of swimming 1 foot per second, an approach velocity of 5 feet per second could be used. The interesting explanation for this is that all fish, as far as we know, on approaching an obstacle in a velocity beyond their swimming ability, will automatically position themselves at right angles to the obstacle. Therefore, by knowing the swimming speed of the smallest fish to be saved, it is possible to determine the required angle of the structure and velocity of approach.

Figure 12 shows the first traveling screen to be constructed. The canal is 6 feet wide, 4 feet deep, and 50 feet long. The screen travels towards

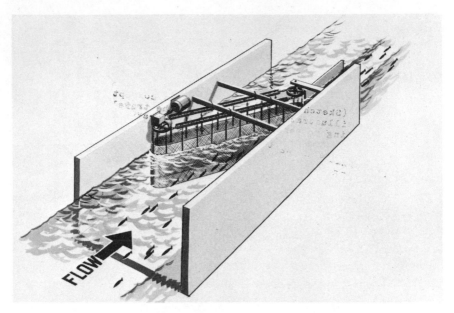

Fig. 12—. Traveling screen No. 1, constructed in 1965.

the bypass, making a turn and then returning upstream through the water. The net extends full depth within the canal. The upstream screen panels are held in a vertical position by two rails located immediately behind the wire-cloth screen.

The advantages of the traveling screen system are economy, efficiency in diverting juvenile migrants, easy deflection of fish, water-depth fluctuation, head loss, the self-cleaning factor, velocity of approach, and simplicity of operation.

Economy.—Since the major portion of the structure under water consists only of relatively inexpensive netting, we believe that considerable savings can be made. For example, we recently completed construction of an experimental traveling screen having a capacity of 1,000 second feet and costing $20 per second foot of water.

Diverting juvenile migrants.—Since fish are unable to pass through the netting, the collection efficiency should be near 100 percent. Mesh size can be changed as required as a factor of fish size.

Easy deflection of fish.—Since the screens are placed on an angle to the direction of flow, fish passing downstream and approaching the structure merely deflect away, moving directly into the bypass.

Water depth fluctuation.—This presents no difficulty, since the facility operates efficiently irrespective of extent or rate of water-depth fluctuation.

Head loss.—With the traveling screen, this is minimal, since we have what is called single-screening, in contrast to double-screening as required for drum screens and industrial water screens. With the traveling screen, the panels, on their return upstream, open up to minimize head loss.

Self-cleaning.—Near the bypass, we arrange for a flow reversal, which washes the screen off from the reverse side. A wash-off system of water jets could be incorporated in the design.

Velocity of approach.—As I mentioned a moment ago, higher than usual approach velocities can be used. Should fish be impinged onto the screen by the force of the velocity, the impingement would be quite gentle, since the travel rate of the screen can be matched to the velocity of approach. The fish and screen blend together. For this same reason, debris impinging gently onto the screen is released readily.

Impinged fish are carried directly into the bypass for release, another very distinct advantage of the traveling-screen system. With all other screens, impinged fish cannot be carried into the bypass and must somehow struggle off the screen or be carried up and over as in the case of drum and industrial water screens. With louvers, fish thrust by velocity onto the structure are washed through and lost.

Simplicity of operation.—All operating parts are up and out of the water for ease of maintenance.

Figure 13 illustrates our latest experimental traveling screen, Model VI. The stiff legs are spaced 10 feet apart. The net panels are not illustrated.

The traveling screen in Figure 14 was recently installed in our Troy test flume on one branch of the Grande Ronde River in northeast Oregon. Water from the river can be diverted into the concrete canal, which measures 143 feet long, 40 feet wide, and 12 feet deep. Hydraulic tests of the traveling screen will be made here shortly.

Figure 15 is an artist's concept of the traveling screen as it would look within a river. Notice the suspension cable system, which was adopted to eliminate costly in-river pier construction. The screen travels downstream across the river with the individual panels fully closed, moves into the bypass section, passes around the end curve, and then returns back upstream with panels open to reduce head.

Figure 16 pictures the louver structure installed several years ago by the Bureau of Reclamation at Tracy, California. The canal has a maximum flow of 5,000 second feet. We hope to design a traveling screen (Model

Fig. 13——. View of an end turn of the experimental traveling screen. This shows
Model No. VI, during construction.

VII) which would be installed within this canal immediately upstream from
the louver structure. At present, the flow passes in this direction through
the trash rack and then through the louvers. The experimental traveling
screen would extend from this point across the canal on about a 25-degree
angle to a point on the far side.

For those of you interested in traveling-screen installations within
powerhouse intakes, there are several ways in which this might be accomplished. For many years, industrial water screens have been placed at right
angles to flow and well within the intake canal. Approach velocities have
generally been set for 1.5 feet per second or less. Fish, on reaching the
screens and being unable to pass through, become trapped within the canal.
Juvenile migrant fish, such as striped bass, shad, salmon, or steelhead

Fig. 14—. The Troy test flume in the
Grande Ronde River, Oregon.

trout, once trapped, would probably never leave the area until forced onto the screen through exhaustion. Mortality would be the direct result of this.

Fish endeavoring to avoid impingement on the screen placed normal to the direction of flow position themselves by heading into the flow. In their search for a bypass, they must deflect to one side or the other. Figure 17 illustrates this. Should they turn 60 degrees to the direction of flow, their swimming speed would have to be increased from 1.5 feet per second to 3 feet per second, to avoid being thrust against the structure. Since this may be physically impossible, due either to limited stamina or ability, impingement and injury could result.

Figures 18 and 19 are diagrammatic and are presented here to provoke your thinking on various concepts for positioning screens within a power-plant intake. Assume that, in the upper part of Figure 18, a river flows in the same direction as our traveling screen. Fish impinged on the screen would be carried down to this point and washed off to continue their downstream movement. In this plan, fish never become trapped within an intake. A trash rack cantilevered out into the river protects the screen structure from damage and through proper design creates no areas for fish entrapment.

Fig. 15.—Artist's concept of the traveling screen as it might appear within a river.

In our second plan, the traveling screen is placed on an angle to flow within the canal. Fish diverted by the screen move into the bypass and from the bypass into a holding pond. Water flowing through the bypass and holding pond is activated by a small pumping plant shown here. From the holding pond, fish can either be released back to the river by gravity flow or collected later for transport by truck to a point distant from the influence of the canal intake. If you are concerned over numbers of fish which might have to be collected and held, it would be helpful to note that the U.S. Bureau of Reclamation each year, at the Tracy, California, louver structure, collects and transports without difficulty over 40,000,000 fish.

Fig. 16——. An aerial view of the Bureau of Reclamation's Tracy, California, louver structure.

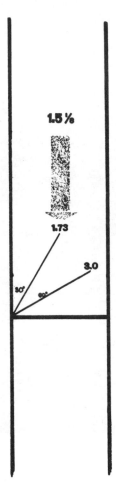

Fig. 17—. Sketch illustrating the problem facing juvenile migrants endeavoring to avoid impingement on a screen placed 90 degrees to flow with a water velocity of 1.5 f.p.s. Fish, on deflecting 60 degrees off from the direction of flow, must maintain a swimming speed of 3 f.p.s. to avoid impingement.

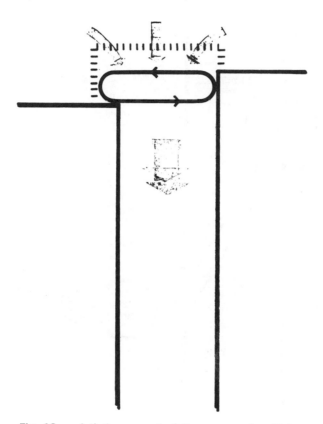

Fig. 18—. Artist's concept of the manner in which a traveling screen could be placed at the entrance to a power plant water intake to avoid entrapment of fish.

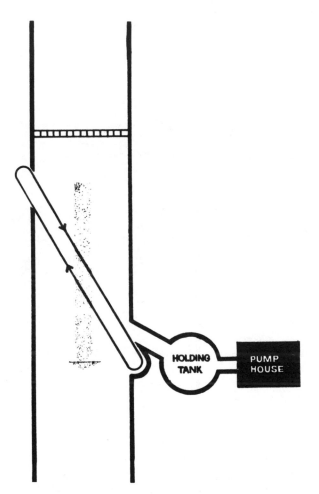

Fig. 19—. Artist's concept of a traveling screen placed on an angle to flow. Fish following movement of water into canal are quickly deflected through the bypass and into the holding tank.

THERMAL POLLUTION

I APPRECIATE the opportunity to be here this evening to talk about the problems of thermal pollution and how we must go about the business of achieving clean water in the nation.

I overheard someone say earlier today: "Joe Moore should be an expert on thermal pollution because he has been in hot water ever since he became Commissioner."

I don't know about the "expert" part of that statement. But considering some of our recent discussions of water-quality standards in the Southeast, there may be a grain of truth in the "hot water" part.

Seriously, I want you to know that we in FWPCA are gratified at the interest shown in these symposia on thermal pollution. As you know, a symposium on the biological aspects of thermal pollution was held in June in Portland, Oregon. If I may paraphrase from an address there by Under Secretary of the Interior David Black, I hope that this meeting on the engineering aspects of heat will produce a great deal of light.

For light—enlightment on the effects of heat in water and how to control thermal pollution—is one of our great needs. Unfortunately, the problems created by thermal pollution have been largely ignored until recent years.

However, since the end of World War II, construction of scores of new power-generating plants—the major users of cooling waters and, thereby, the great dischargers of heat into the environment—has resulted in a growing public concern about hot water. This sudden rise in public awareness has far outstripped our knowledge of the ecological effects of large discharges of heated water in our rivers, lakes, and estuaries. We also are sadly lagging in discovering and designing better and more economical

243

methods of controlling heat discharges for the greatest variety of beneficial uses.

A public that is attuned, as never before, to the needs of preserving our environment seems to demand that we discover and implement ways of saving our waters not only from heat but from all the other pollutants that defile them.

Certainly, curtailment of power generation is not the solution to controlling thermal pollution. If this nation and this region are to continue to prosper, we must have more and more electric power and more power-generating plants. I need not tell you, here in the heart of TVA country, that economic well-being is dependent on adequate power generation.

It is estimated that by 1978—just 10 years from now—a volume equal to approximately one-sixth of the total freshwater runoff in the United States will be used for cooling and condensing purposes. This, of course, does not include the cooling water that will be used by nuclear power plants that are being built on coastal waters throughout the country.

Right now, fossil-fuel steam-electrical generating stations are the greatest users of water for cooling, accounting for 95 percent of the total production. But we already are in the first years of a new era that will see giant nuclear reactors replacing the conventional plants at a rapid rate. By 1980, nuclear plants will have taken over 35 percent of the job of making electricity. Now, nuclear plants account for only five percent of thermally generated power.

While nuclear plants can turn out more power more economically, they will produce 40 percent more waste heat than coal-fired plants of the same size.

In addition to the greater volume of cooling water required, nuclear power plants will add 30 percent higher temperatures to cooling waters than do present conventional plants.

But I am talking as if the age of nuclear power generation and the age of grappling with great problems of thermal pollution were still a decade away. Unfortunately, we face many of these thermal problems here and now. They must be dealt with in the immediate future.

Let me give you some examples right here in the Southeast:

TVA has recently received a permit from AEC to construct a nuclear power facility at Browns Ferry, Alabama, on the Wheeler Reservoir of the Tennessee River. The cooling-water requirements for this plant will be an astounding 2.8 billion gallons per day. The temperature increase in the cooling-water, from intake to discharge, will zoom an estimated 25 degrees Fahrenheit in a matter of about six minutes.

According to TVA's proposal, this heated water would be discharged through a diffuser pipe laid along the bottom of the reservoir channel in order that the heat load may be mixed and diluted as much as possible by the receiving waters of the reservoir. However, even with this controlled discharge, TVA expects that there will be a continual 10-degree-Fahrenheit increase in temperature in the vicinity of the diffuser. Can we really say— based on our present knowledge of thermal effects—that this temperature increase will not have harmful effects on present beneficial uses of the reservoir, such as recreation and fishing? There is no question that this additional electric power will be needed to bolster the continuing prosperity of the Tennessee River Valley. But we must be careful that, in seeking greater economic prosperity, we do not irreparably damage our environment.

The Browns Ferry Project is, I believe, a good illustration of possible problems created by a plant constructed on a freshwater site. But what about the nuclear plants that are being built on the nation's estuaries and coastal waters? If we approach thermal discharges into fresh water with fear and trepidation, how should we treat proposals to use our fertile estuarine waters for condenser cooling? For purposes of illustration, say that our knowledge of the effects of hot-water discharges into freshwater resources has a worth of, say, one dollar. Then, by that standard, our knowledge of estuarine effects would be worth about ten cents. And I am not sure this is a total exaggeration.

Decisions on constructing these much-needed coastal power plants cannot wait for an explanation of all the complexities of the effects. In Florida, right now, construction is proceeding on two of these mammoth facilities, one on the Gulf at Crystal River and the other at Turkey Point on Biscayne Bay.

Let us talk for a moment about the Turkey Point facility, with which you may already be familiar. Two fossil-fuel generating units already are in operation at Turkey Point. These installations produce a maximum capacity of 740 megawatts and pour through their condensers 192 million gallons of cooling water every day. By 1972, with the installation of two nuclear units at the same site, the maximum capacity of the plant will be approximately 2,200 megawatts, utilizing 3.15 billion gallons of cooling water each day.

Our FWPCA Southeast regional pollution surveillance team tells us there is already evidence of a substantial temperature increase in Biscayne Bay with the present fossil-fuel cooling-water discharge. What will be the result when the cooling-water volume is increased more than 16-fold?

Biscayne Bay is a national treasure. It is a unique, subtropical manifestation of God's grand design for maintaining equilibrium in nature. It has been proposed that a portion of this great ecological resource be set aside as a national monument for the enjoyment of generations to come.

Certainly, the owners of these generating facilities at Biscayne Bay have exhibited great interest in preserving much of the natural beauty of this area. They have demonstrated, time and again, their concern for preserving the area, lest they be known as nature's spoilers instead of civilization's benefactors.

Can we give them rational guidelines for protecting this environment while still allowing them to fulfill their obligation of servicing and stimulating the economic health of the community? We must seek some meaningful answers, not only for Turkey Point, but for similar facilities across the nation.

First of all, it seems to me, we must consider methods of controlling and managing waste heat in the aquatic environment.

Cooling towers already have been demonstrated to be an effective method of control. However, the cost of these enormous structures can add between $5 and $10 per kilowatt to the cost of plant construction, and, therefore, the towers become a significant factor in considering the economic feasibility of the entire installation.

There are other alternatives:

—Dilution of waste heat through co-ordination with releases from storage reservoirs, at the same time that cooling waters are being released from power plants.

—Use of aerated ponds to facilitate the exchange of heat to the atmosphere before release to receiving waters.

—Recycling of condenser water, thereby reducing the volume required and minimizing the wasteful effects of the once-through process now employed. And perhaps we might even consider another approach:

—Closing down the plant or operating at minimal level during periods when the addition of heat would be most hazardous.

But as the need for more power production continues to accelerate, we must take a long-range planning view of how to deal with thermal pollution.

We must carefully consider the proposed location of each plant so that its impact on the environment will be most beneficial. Waste heat from a power-generating plant situated on productive and ecologically sensitive receiving waters becomes "a resource out of place" and therefore its economic benefits may be more than offset by the harm it does to the environment.

The future will demand that, beyond simple planning, we must improve the efficiency of our present systems and develop new means of generating power. We must seek out ways to harness the heat we now aimlessly and destructively discharge to environments.

Eventually, we must develop accurate predictive methods that will permit the logical setting of water-quality standards based on effects rather than the use of relatively arbitrary temperature figures.

Because heat affects every organic factor in determining water quality and, therefore, water use, the problems of dealing with thermal pollution are among the top priorities of the Federal Water Pollution Control Administration.

The spiraling needs of population and industry for more electric power make it essential that we move forward in devising cohesive management options for dealing with waste heat.

As most of you know, FWPCA has already initiated a long-term study of temperature effects on the Columbia River which we hope will give us some of the answers. But this problem is too complex and too widespread to give us total solutions from a single study.

As an example of cooperation with the academic field, research grants have been made to Vanderbilt University for a thorough review of the literature on the thermal problem. The University of Miami Institute of Marine Sciences also has received an FWPCA research grant to study the long-term interrelation among aquatic species in Biscayne Bay and their reaction to increased heat. In addition, FWPCA is actively considering a technical assistance program in the Biscayne Bay area which will, in part, provide some immediate answers to problems of most effectively controlling the heated effluent from the Turkey Point Plant.

Needless to say, these projects cost money, lots of money. Someone has to pay the bill. I cannot emphasize too strongly that we must not be misled into thinking that we can have clean water—and, in this case, cool water—on the cheap. We hear a great deal about whether municipalities, industry, the states, or the federal government should pick up the tab for water-pollution control. Talk about who should pay for what begs the question.

The bills for removing waste from our waters—whether that waste be municipal sewage or destructive heat—are paid by people. We can talk for days about cleaning and cooling water, but we must be willing to pay for what we want. And I can promise you the cost of clean water will not come down. While someone else may temporarily absorb the cost of keeping the water free of excess filth, as well as heat, ultimately the money will come out of our pockets. That cost may be in the form of taxes or additional charges on our utility bills or it may be borne in the cost of products which we consume or utilize.

It is futile to insist that our national waters be kept clean and at the same time argue that the other fellow should pay the bill.

The cost of clean, cool water must become merely a part of the expense of doing business in this country. If we are not willing to assume that cost, then we are really not interested in clean water.

But as I have said, the public has become aware of and more concerned about destruction of the environment by uncontrolled pollution. Because this public awareness has become so evident and so dramatic in recent years, I personally believe the American people are willing to pay the bill for controlling the contaminates and pollutants which now threaten our prosperity and our way of life.

Chapter 9   Charles Waselkow

DESIGN AND OPERATION
OF COOLING TOWERS

WELCOME to the Cooling-Tower Club! And, I might add, it's high time you did join the club. You have too long enjoyed freely the bountiful gifts of nature insofar as rivers, lakes, and oceans are concerned!

May I hasten to admit that this outburst stems from envy. I am sure you will find it in your hearts to forgive us Westerners where the subject of water is concerned—water in quantities required by our power-generation industry!

But first, let me state from the outset that I am not posing as a cooling-tower design expert. I understand that, due to the rapid-transportation facilities available today, the definition of expert being so many miles away from home has increased considerably, so that I am eliminated from any competition with real experts in this field.

I do, however, live in a part of the country which, in spite of being a wonderful place to live, does require that we conserve our water resource to the utmost. The cooling tower is one of the resource-conservation phases of our daily life.

I would also like to limit my subject for today—"Design and Operation of Cooling Towers"—to deal more with design criteria associated with the selection of a cooling tower as a result of experiences, rather than the theoretical design details of the tower structures and evaporation theories.

I would like to give you a brief recount of the cooling-tower history on which I will base my remarks. These remarks will include the tower experiences of Public Service Company of Colorado; Southwestern Public Service Company of Amarillo, Texas; Utah Power and Light Company; and the consulting-engineering firm of Stearns-Roger, of Denver, Colorado.

Fig. 1—. Public Service Company of Colorado, Cherokee Unit No. 4 cooling tower, 180,000 gpm., 350-mw unit.

(Photograph by Public Service Company of Colorado)

The first cooling tower on record in Public Service Company of Colorado was installed at Leadville, Colorado, in 1907, and was used in conjunction with a vertical turbine-generator of 1500 kilowatts.

Today, we are operating twelve cooling towers for a total of 61.8 percent of our total system-generating capability, and 76 percent of our steam-generating capability.

It may interest you to know that, at the design stages of our latest 350,000-kw steam-electric unit, at the Cherokee Station of Public Service Company of Colorado, it was found to be the largest tower at that time which required new design criteria to support the distribution piping system. Its dimensions are 489 feet long, 71 feet wide, and 35 feet high (Figure 1 and Figure 2).

Southwestern Public Service Company of Amarillo has indeed had many

Fig. 2——. Public Service Company of Colorado, Cherokee Unit No. 4 cooling tower, 180,000 gpm., 350-mw unit.

(Photograph by Public Service Company of Colorado)

years of experience with cooling towers, since rivers and lakes are not too plentiful in that part of the country. The annual evaporation of 60 to 84 inches per year results in man-made lakes being rather wasteful of water. In Colorado, we estimate that the water requirement by lake-cooling per kilowatt of installed capacity is about 50 percent more than in a closed circulating cooling-tower system.

At the present time, Southwestern operates some 35 cooling towers, ranging in capacity from 6000 gpm to 146,000 gpm. The principle vendors represented are Marley, Fluor, Pritchard, and Lilie-Hofman. These towers range in age from one year to 28 years. The older towers are all redwood, while the later units are of treated Douglas fir. The latest towers use treated fir structural framing and fiber glass outer walls. The fills are plastic, using both polyethylene and polypropylene. The fan stacks of the largest and newest towers are made of glass-fiber-reinforced polyester plastic, impregnated with their standard turquoise pigment. The specs call for silicon bronze nails and hardware throughout, and stainless steel pipe banding where wood-stave piping is utilized. Fiber glass distribution piping economics are now under study.

A word of caution at this point may be in order, in regard to the design of the saddle supports under the large wooden-stave distribution piping: for smaller wooden-stave piping, the usual partial-arc saddle was quite adequate; but the large sizes, such as that required for a 150,000-kw unit and larger—these supports require a good deal more design effort. The partial-arc supports actually cause the stave pipe to collapse, requiring complete re-assembly of the whole section of pipe.

COOLING TOWERS

The cooling-tower field is one about which the least of a definite nature is known. The very process of cooling is complicated and becomes more so by the fact that cooling towers are also subject to the whims of the weather. I believe it's a remarkable feat for a tower manufacturer to be able to predict and approach the performance stipulated in a given specification as well as he does, without the use of ridiculously large factors of safety.

The cooling tower constitutes one of the most difficult items of plant equipment to specify and evaluate. My predecessor in our Mechanical Engineering Department devoted over 20 years of his work-life to studying cooling towers and lake cooling. In fact, his lake-cooling ASME paper and formuli are recognized by many as a basis for determination of lake-cooling designs of this day.

I remember one of his greatest disappointments was the fact that the last cooling tower which he so painstakingly specified, after all those years of study, could not be proven as performing up to his specification.

So, in addition to this magnitude of uncertainty in the cooling-tower design, a controversy over whether a cross-flow cooling tower evaluated more or less favorably than a counter-flow type raged for a long time. My predecessor was firmly convinced that the cross-flow, induced-draft type resulted in a more economical tower—especially on the input power to the fans.

It seems his convictions were recently confirmed by statements from qualified people in the industry that the power input is lower in the cross-flow tower.

W. E. Robinson, vice-president of Southwestern Public Service Company, states that they have purchased both cross-flow and counter-flow towers; but their latest purchases have been cross-flow design, which, in some cases, can effect a substantial saving in the concrete-basin costs. As an example, their last purchase of a 146,000-gpm cross-flow tower showed a saving of over 10 percent in basin costs over a parallel-flow tower of the same capacity.

Since a cooling tower is at the mercy of the whims of the weather, one of the truly critical design criteria necessary to a good cooling-tower design is the determination of the design inlet wet-bulb temperature. This duty cannot be relegated to the computer. This important item is entirely a local, on-site, proper daytime ambient wet-bulb temperature determination. This wet-bulb temperature should include the weather conditions encountered during the four summer months of the year.

Due to the fluctuations in weather from year to year, care should be exercised not to base the design on the weather conditions of a given year which may not be representative of the weather conditions in other years.

Any company being faced with the possibility of using wet-cooling towers in the foreseeable future must by all means compile its own set of weather conditions in order to be able to specify intelligently the wet-bulb temperature required by the specification for a cooling tower. Dry towers require a history of dry-bulb conditions. Each degree represents real money costs in this important area of power-plant equipment, so that not only must the wet-bulb numerical values be considered, but their frequency or duration during a year should govern the selection of a wet-bulb figure for a given tower. The same, of course, is true for the dry-cooling towers. One robin does not make a spring: neither does one observed wet-bulb or dry-bulb temperature constitute a design criteria.

As a guide for those who may be interested in specification figures, the newest wet towers purchased by Southwestern Public Service Company and Public Service Company of Colorado are as follows:

TABLE 1. Specification figures: wet towers, Texas and Colorado.

Southwestern Public Service Co.		Public Service Co. of Colorado	
Hot Water	93.8° F	Hot Water	98.1° F
Cold Water	79.1° F	Cold Water	80.0° F
Wet Bulb	70.0° F	Wet Bulb	60.0° F
Drift Loss	0.1%	Drift Loss	0.1%

COOLING WATER

Indeed, fortunate are those among us in the power industry who possess the water resource and are only confronted with a thermal-pollution problem! In this case, a cooling tower can cool and aereate the condenser water sufficiently to be acceptable to the thermal-pollution authorities.

But reflect for a moment on the dilemma of a design engineer who is confronted with insufficient make-up water to a cooling-tower system and thermal stream-pollution at the same time! This is one of our problems.

Conventional steam-power plants have used seawater, rivers, or lakes as a source of cold water to dispose of the heat rejected to their condensers. In locations where these sources are unavailable or inadequate, cooling towers or spray ponds are used to reject this heat to the atmosphere by evaporation. With the growth of the steam-electric generating industry, the economic sources of cooling-water supply are becoming exhausted. There has been much activity in this field of water resources, their conservation and utilization.

In order to get a feel for the quantities involved in this area of our activities, I would like to quote a few figures regarding the cooling-water requirements.

The most efficient steam plant rejects about 4500 BTU per kwhr to its condenser, while the average steam plant rejects about 5500 to 6000 BTU per kwhr. In the case of a cooling tower, this is an actual consumptive use of from one-half to three-quarters of a gallon of water per kilowatt hour. For nuclear plants of the boiling-water or pressurized-water type, the comparative heat rejection is about 7250 BTU, or about 0.9 gallon per kilowatt hour consumptive use. As a rough proportion, a coal-fired plant will use at least five pounds of water per pound of coal.

The location of conventional steam-electric generating plants is determined by availability of cooling water, among other items. Many plants accept sizeable transportation costs of fuel and generated energy in order to obtain adequate cooling water.

At Southwestern Public Service Company, they have established the use of sewage effluent for cooling-tower make-up. Mr. Robinson states they have gone to this make-up source for three reasons: (1) Its use conserves fresh ground-water and thus becomes a conservation measure, a policy adhered to for many years. (2) Power requirements increase with a city's growth, and growth of a city makes available increasing amounts of industrial water. Thus, the water supply becomes firm for the life of the plant. (3) Revenue derived by the city through sale of its industrial water provides a means of expanding sewage-treating facilities without large cost to the citizens. This is a pleasant experience for the taxpayers and improves the image of the utility involved.

Southwestern's experience in the use of sewage effluent goes back to 1960. Mr. Robinson points out that the water required by the first units installed is not particularly cheap, since rather extensive supplementary treatment with cold-lime treaters is required to precipitate silica, magnesium, and some calcium. However, with long-term contracts, the water becomes more economical with each succeeding unit installed. This water, containing a maximum of 25 ppm BOD (Biological Oxygen Demand) and 25 ppm of suspended solids, has given no cooling-tower problems, although it is found that the drift is stickier than normal with fresh water. Foaming has not been objectionable.

DRIFT LOSS

The matter of drift loss must be taken into consideration for several reasons. It is virtually impossible to have a power station cooling tower which will not involve the electrical transmission lines and switching equipment.

In Public Service Company of Colorado, we experienced main-transmission-line flash-over several times, in spite of efforts to wash the insulators periodically. The eventual solution was the physical relocation of the entire transmission line away from the cooling towers and the establishment of a rule that future transmission lines cannot approach any cooling towers nearer than 500 feet. This rule has been observed for several years now, and the 500-foot clearance appears to be the correct one for Public Service Company of Colorado water-quality conditions. Unfortunately, this type of solution is not always available to the power-station layout engineer.

At Southwestern Public Service Company, the question of drift loss is even more critical. You will recall that they use sewage effluent for cooling-tower water at their Nichols Station.

For many years, they had purchased cooling towers based on the usual 0.2 percent drift loss as standard specification. In general, this has been satisfactory, but the installation at the Nichols Station at Amarillo in 1960 and 1962 proved that 0.2 percent drift loss is excessive. The sewage effluent contains detergents and some organic matter which proved troublesome when the drift spray was carried to the nearby substation by the prevailing winds. This spray produced a very tenacious coating on the 115 kv bushings and insulators, resulting in frequent flash-overs. During the first three years of operation, the insulators had to be cleaned at three-month intervals. Washing with a high-pressure jet of condensate was effective in removing most of the deposit, but unfortunately, a serious flash-over with the loss of a bushing on the auxiliary power transformer resulted, and hand cleaning had to be instituted.

One of the remedial measures taken to improve this condition was the installation of additional mist eliminators in the plenum of the cooling tower. Tests determined the drift loss from fan stacks had been reduced approximately 0.1 percent and this reduction lengthened the insulator-cleaning period to once a year, instead of four times per year.

The other helpful measure instituted by Southwestern was the use of silicone grease as a coating for the insulators. The grease is applied by hand and wiped off and reapplied at yearly intervals.

Public Service Company's experience in this area did not seem to result in an appreciable improvement in the drift-loss control. On our eight-cell cooling tower on a 100,000-kw unit, the drift eliminators were an afterthought. The entire tower was completely installed, but the main unit had not yet been activated when the decision to add eliminators was reached. This required a considerable amount of alteration and fill removal. Anyone in need of redwood kindling, please contact me!

At the same time, it was decided to modify only seven cells, out of a total of eight, so that the drift losses in each type of design could be determined under identical operating conditions. On paper, this was a test man's Utopia.

The tests were carried on by a very competent professional talent. The most discouraging part of all of this effort was the statement in the final report: No appreciable drift-loss difference exists between the original cell and the adjacent cells equipped with the new drift eliminators! This, too, was a blow to my predecessor.

In spite of Public Service Company's experiences in this area, drift eliminators should be included in all towers, and their performance specified to such a figure as 0.1 percent. With new regulations governing air-pollution being formulated almost daily, injecting any substance into the air which may become harmful to humans, agriculture, or buildings and equipment is likely to be prohibited in the near future.

In fact, in our own Denver area, we have one individual who, for the past several years, has "officially" complained that we are "not fooling" him. "That white steam-like stuff" coming from one of our cooling towers near his home "is loaded with radioactive material". He "knows" we are "making atomic stuff on the Q.T."

There is one for our public relations department!

OPERATION OF COOLING TOWERS

For those of you who may not as yet have had the experience of operating cooling towers, may I say that you have nothing to fear! It is one of the better ways to perform the function of condensing our turbine steam. In Public Service Company of Colorado, we have as varied a field of experience as one needs to evaluate the merits of the various possibilities.

We have a 45-year recorded history of lake cooling. We have plants operating on 100 percent stream flow; we have plants utilizing partial cooling-tower river-flow; and we have three of our newest plants and towers operating on 100 percent cooling-tower cycles.

Lake cooling is by far the most trouble-free type of operation we have, except for the need for more make-up water to the lake, which is not too readily available in this part of the country (Figure 3).

The second choice is cooling towers. The last and most undesirable type, in our experience, is the once-through stream-flow cycle.

These choices are based primarily on the operating viewpoint. That is, how much of a problem to the operating personnel is each type?

With lake cooling, we have no problem except to keep enough water in the lake each year to permit the operation of the plant.

The cooling-tower operation, of course, requires more attention. However, it is dependable. Its performance is predictable and is relatively constant over the various seasons of the year. It does require close supervision over the lubrication system. It does require close surveillance during icing conditions on the inlet louvers. This is only during the few winter months. The icing condition is routinely handled by: (1) the manufacturers' design of suitable inclining-inlet surfaces, (2) by the simple expedient of reversing the direction of the cooling-tower fans and defrosting the ice formations.

In the past, we resorted to the use of two-speed motors, with the objective of using the slow speed in the reverse direction for defrosting purposes. This practice has been found to be an unnecessary investment. All our newest cooling towers are now equipped with dual-speed fan drives. The forward and reverse speeds are equal. The manufacturers have improved their lubricating and bearing designs to permit full-speed operation in either direction without damage to the gearbox components.

Fig. 3——. Public Service Company of Colorado, Valmont Station, showing two of three man-made cooling lakes.

Reprinted by permission, from **Lines** 10 (1964), Public Service Company of Colorado. Photograph by Birlauf & Steen, Inc., Denver.

One very important item, in regard to the operation of the fans in reverse for de-icing purposes, that must be provided to prevent damage to the gearbox drive-shaft during reversal is a reliable automatic brake. This applies whether the fan is equipped with a slow-speed provision, or is merely coasting, or is being driven by the natural-stack draft action in the tower structure. We experienced years of shaft damage on our older models before we installed home-made automobile hydraulic brakes on each fan.

In connection with icing, I would like to have you visualize the sensation a station supervisor experiences when he is called out at three o'clock in the morning to look at a cooling tower which is one solid cubicle of ice! This is one problem that is seldom discussed; yet, every utility will experience it sooner or later, since it is present in every winter life of a cooling tower. Every time a unit is removed from service during winter months, there is present a possibility that the circulating-water system will not be shut down until it's too late! Being an independent system, the circulating-water system is easy to forget, if the shut-down is an abnormal one, in the first place.

May I pose a question? How would you go about saving a cooling tower if you found you were in charge and you had this situation on your hands? Having lived through this experience and having been able to save the tower fill, I want to say that the first impulse you may have to start the circulating pumps must be resisted. Remember, the ice would melt from the bottom up, and the fill would be subjected to the total weight of the ice. In my case, the procedure that resulted in saving the fill was to melt the ice slowly by hosing the sides of the ice until the total volume was considerably reduced.

COOLING-TOWER WOOD DETERIORATION

The experience with wood deterioration on the Public Service Company towers is interesting, as well as being unexplainable. In order to explain some of our problems with wood deterioration, and why it is puzzling us, it should be pointed out that three of our major plants are located from south to north along the Platte River through the heart of Denver.

All three plants use substantially the same cooling water for their circulating-water systems.

Redwood is common to all three stations, even though Douglas fir is predominant in the newer installations. Yet the deterioration of the redwood at any one of the three stations bears no resemblance to that at either of the others.

One station has a life-history for its redwood in the order of five years with two complete sets of fill-replacement in the past ten years. The other two stations have practically no indication of redwood deterioration, to date; and, of course, there is no noticeable deterioration of the adjacent treated-Douglas fir towers. The ages of these fills are considered here on a comparable basis.

At first glance, one may be tempted to explain this behavior of the tower fill on the most obvious, improper chemical treatment, etc. In this respect, may I point out that the same chemist supervises all three stations personally! All equipment and modes of treatment, etc., have been checked and cross-checked for years, to determine whether operating peculiarities existed which could be contributing to the failure of three towers in one plant and none in the other two stations.

In addition, all operating personnel bid and flow between all three plants as if they were under one roof, so that the same personnel sooner or later carries on similar duties in any plant.

The only physical differences between the station experiencing the fill difficulties, as against the other redwood installation in the other power stations, is the fact that the circulating-water system is common to all four units in the station, whereas the other two stations have unitized systems. Make-up water to all three stations is, of course, the Platte River water (Figure 4).

To date, no explanation has been found for this behavior. After the initial failure, which occurred five years after its initial operation, we replaced the untreated redwood with a treated redwood. After the second failure of the fill, this time the treated type, we replaced the redwood with treated Douglas fir. The behavior of the Douglas fir cannot be reported today, since it has been in service for one year only.

We attempted to relate this behavior to certain algae and bacteria which may be present in the Platte River at the south end of the Denver city limits, but this, too, did not exist.

Furthermore, we provided deck coverings for a few cells, in order to observe the effect of sunlight on the algae formations. We found that, with a common chlorine treatment of the circulating water, there was no observable difference between the shaded decks and the unshaded.

Our only hope for the towers in this particular plant is that the new Douglas-fir fill will provide at least a partial solution to this puzzling behavior of our tower fills.

PUBLIC RELATIONS

With the public awareness of air pollution and thermal stream pollution, the pressure on the station-design engineer has become increasingly greater.

Fig. 4——. Nichols Station of Southwestern Public Service Company, showing multi-tower installation on common tunnels.

(Photograph by Birlauf & Steen, Inc., Denver)

Often, his sense of evaluated worth is jarred by the amount of money that is necessary to produce an acceptable situation in the realm of public relations. An aroused public can make demands which are often beyond the realm of reality. I have, on several occasions, been called by people in highly placed ranks of our society, who simply ordered me to "shut that plant down!"

Today, we are in the process of installing a $100,000 settling pond, just to handle a small amount of carry-over fly-ash from one of our smaller stations.

May I again point out that the stream pollution, out West, is, again, a different problem from what it may be, back East!

The Platte River is NOT an Ohio or a Hudson river. On occasions, I have seen flows of 4000-GPM total stream flow. Unfortunately, I have also

Fig. 5—. Public Service Company of Colorado, Zuni Station, showing debris and damage to the No. 1 and No. 2 main cooling towers after the flood of 1965.

(Photograph by Birlauf & Steen, Inc., Denver)

witnessed a 50,000-second-foot flow in this same stream, that filled two of our major stations to a depth of 13 feet of water in one, and 6 feet in the other (Figure 5 and Figure 6)!

In a situation such as ours, it takes very few BTU input to the stream to affect the total temperature of the stream and the allowable BOD (Biological Oxygen Demand).

Cooling towers are certainly a practical solution to this problem. A blow-down rate of 240 GMP for a 100,000-kw station employing an 89,000-GPM circulating-water system results in very little thermal stream-pollution and especially if the blow-down water is re-used for ash-removal purposes, as in our major 750,000-kilowatt station.

One other item of public relations, as it concerns cooling-tower operations, is the problem of road hazard. As I stated before, our major stations are located almost diametrically across the city of Denver, along the Platte River. One can readily visualize how the cooling-tower vapor can become a problem on the adjacent streets and highways, if the atmospheric conditions are right and the construction of the tower is not proper.

Fig. 6—. Public Service Company of Colorado, Zuni Station, showing undamaged service water cooling tower after flood of 1965.

(Photograph by Birlauf & Steen, Inc., Denver)

On one occasion, I was horrified to witness cars plunging into the steam-cloud from one of our older cooling towers and to hear them crashing against each other, somewhere in the vapor!

Since that time, all of our towers are designed considerably higher in structure and all are equipped with venturized fan-discharge cylinders, which permit the discharge of the vapor higher into the atmosphere.

FUTURE OUTLOOK

With the natural sources of water from the seas, rivers, and lakes becoming more and more limited, either by lack of quantity as a result of its greater use by man, or through *man-made* restrictions due to thermal pollution, new methods must be found to cope with this serious problem. I would like to mention a few approaches to this problem, which, today, are in various stages of development.

Fortunately for mankind, there are alternate heat sinks, such as air, to

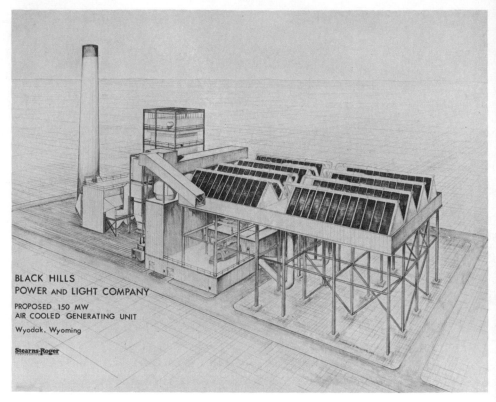

Fig. 7——. Black Hills Power and Light Company proposed unit at Wyodak, Wyoming.

(Photograph by Birlauf & Steen, Inc., Denver)

dissipate unuseable heat wastes. The know-how is here; the only obstacle is money! How different is this problem from those of the Pharaohs of ancient times, who had "money," labor, etc., but lacked the know-how to resolve some of their engineering problems!

Except for this item known as *money,* the engineer of today can "take to the air." He can resort to cooling lakes, spray ponds, hyperbolic (wet or dry) natural-draft towers, induced-draft towers, etc.

In addition, he can resort to Air-Cooled Condenser of the GEA System (Figure 7).

And finally, some day, he hopes to convert heat energy directly to electrical energy with a minimum requirement of water and a minimum of air pollution and water pollution.

I am sure you are all familiar with the hyperbolic natural-draft towers

For six centuries, Cannock Chase has provided fuel and power for the service of man . . .

Fig. 8——. Illustrates the relative size of wet and dry types of natural-draft cooling towers of equal ratings.

(Photograph by Birlauf & Steen, Inc., Denver)

which are being introduced into the power industry at this time. The wet natural-draft cooling tower is gaining in popularity as a corrective thermal-pollution measure being taken today, even though it still consumes water at the same rate as a conventional cooling tower.

The day of dry towers on a large scale is yet to come. The dry tower's water consumption is very low and its affect on stream thermal-pollution is nil (Figure 8, Figure 9, and Figure 10).

The wet hyperbolic natural-draft towers now in practical use in suitable climatic areas do solve the thermal pollution of adjacent streams, but their appearance is very much against them. This in itself would limit the number of available station sites. A structure of 30 stories high and some two city blocks in diameter just cannot meet any one's sense of esthetic beauty. The recent Fort Martin Station of the Allegheny Power system has the largest in the world, at 370 feet high and 380 feet in diameter to serve one 540-mw unit (Figure 11).

The cost of such structures makes them difficult to justify, economically. A wet natural-draft tower for a 1000-mw unit may run as much as $10

Fig. 9——. Dry type, natural-draft cooling tower.

(Photograph by Birlauf & Steen, Inc., Denver)

per kw. In addition, even if water for make-up were plentiful, the land values may become prohibitive, inasmuch as a 1000-mw station may require 1000 to 2000 acres of land!

A rather unusual cooling-tower design suitable for unusually water-scarce areas is the GEA Air Exchangers, Inc., Bochum, West Germany,

Fig. 10——. Fifty-foot coils of dry type of natural-draft cooling towers.

(Photograph by Birlauf & Steen, Inc., Denver)

who have numerous such installations in service in Europe (Figures 12 and 13).

The GEA Air-Cooled System conducts the exhaust steam from the turbine, through large ducting, directly to a header, from which two inclined

Fig. 11——. Artist's conception of possible future plant and cooling-tower
arrangement.

(Photograph by Birlauf & Steen, Inc., Denver)

sections of air-cooled finned-tube surfaces are fed. Air is supplied to the
center of this inverted-V arrangement by large fans similar to conventional
cooling-tower fans. The headers at the lower ends of these condensing sections feed additional after-cooling sections, where the remaining steam and
noncondensables flow upward. These sections are provided with separate
fans and proportioning of duty between condensing and after-cooling sections is accomplished by adjustment of fan speeds. This minimizes power
consumption and prevents freezing problems during cold weather or light-

Fig. 12——. Typical GEA installation.

(Photograph by Birlauf & Steen, Inc., Denver)

load operation. This patented arrangement also avoids subcooling of the condensate which is drawn off the connecting header, between the condensing and after-cooling sections.

One of the first installations of the GEA systems was installed in Europe in 1939. The investment cost of such a plant is considered to be comparable to that of a conventional plant, because the low-water consumption permits the plant to be located at the mine mouth. The additional cost of the air-cooled condenser installation is offset by the savings due to two principal features: first, as a mine-mouth plant, no coal unloading or preparation facilities are required. The mine will deliver sized coal to the plant reclaim area, as required. Second, with the air-cooled condenser, no condensing water supply, storage, or treating facilities are required.

In summary, there are definite potential applications for the use of dry cooling or direct air-cooled condensing facilities to utilize the low-cost fuel

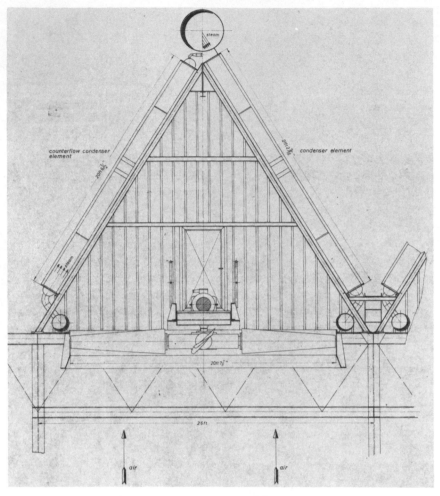

Fig. 13——. End-view drawing of GEA air condenser.

(Photograph by Birlauf & Steen, Inc., Denver)

reserves and, above all, to conserve the limited water resources for more critical uses.

CONCLUSION

In conclusion, may I state that, fortunately for us, practical solutions exist today for the prevention or minimizing of thermal-pollution, as well as conservation of water resources.

The nuclear-power-generation field has attracted considerable attention

Fig. 14——. Public Service Company of Colorado, Fort St. Vrain Nuclear Generating Station.

(Photograph by Birlauf & Steen, Inc., Denver)

to itself by virtue of the fact that some present-day cycles release as much as 50 percent more waste heat per kilowatt of electricity generated than do fossil-fuel plants, partly because the light-water nuclear plants are, from a heat-utilization standpoint, less efficient than the fossil-fuel plants.

Here, too, we have a practical solution. You may recall that we in Public Service Company of Colorado are in the process of installing the first commercial HTGR (High Temperature Gas-Cooled Reactor) in the U.S.A., at our Fort St. Vrain Nuclear Station in Colorado (Figure 14).

This cycle utilizes a 2400-psi pressure and 1000° F. main steam and 1000° F. reheat, similar to our present-day fossil-fuel plants.

Its cycle efficiency, however, approaches 40 percent: the end result of which is, of course, less waste heat to the stream than even from a fossil-fuel plant!

Gentlemen, it would seem that the answers to thermal pollution are available today. All that is needed is MONEY!

REFERENCES

Parse, J. Y. 1967. "Cooling Water Scarce? Use Air!" *Electric Light and Power* (11), Power Division, Stearns-Roger Corporation, Denver, Colorado.
Public Service Magazine. 1968. "Is There Thermal Water Pollution?" (June)
Robinson, W. E. Personal communication.

DISCUSSION/ Gordon L. Ford

WHAT you have just heard is a very good presentation of the anxieties and problems of one who has had considerable experience with a particular type of cooling tower with a particular type of problem. I think what makes life interesting is the fact that we do not all have the same problems; therefore, we do not always have a ready answer when problems do present themselves.

Now that I have mentioned that we do have problems, perhaps I should add that the really important thing is to recognize their magnitude. For example, we are met here primarily to discuss thermal pollution; but, as is evidenced by the fact that the states and FWPCA cannot yet agree on a temperature criterion for streams, the magnitude of the problem is unknown —but more about this, later.

Mr. Waselkow pointed out that he was not posing as a cooling-tower design expert; neither am I. But we both have experience in establishing concept, design criteria, and ultimate operation. You see, the utility decides the conceptual operation, provides the operational data; but the cooling-tower contractor actually sizes, designs, constructs, and guarantees operation to those criteria. Until about one month ago, TVA was completely inexperienced in the operation of cooling towers.

At our Paradise Steam Plant, two units have been operating at nearly full load for one month on one cooling tower which was never intended to operate in this manner.

Mr. Waselkow marvels in the ability of the cooling-tower designer and manufacturer to be able to predict and approach the performance stipulated in a specification without the use of a ridiculously large factor of safety. I read just the other day that, a decade ago, including a safety factor of a "little 3° F." was the common practice. Quite often, the little 3° F. turned out to be a 50 percent safety factor, as far as cost was concerned.

Material appearing here on pp. 276–279 appeared previously in *Power Engineering,* 71(1) January 1967.

For instance, a given tower, to cool 28,500 gpm from 118° F. to 88° F. with an 80° F. wet bulb would cost about $6.10 per gpm. But a tower to cool the same water from 115° F. to 85° F. at 80° F., wet bulb, would cost $9.00 per gpm (Maze, 1967). So you can see that, in competitive bidding, as we have today, the engineer must take real care in his design.

Mr. Waselkow states also that a controversy exists over whether a cross-flow tower or a counter-flow tower would evaluate more favorably. At this point, I would like to throw in some more controversy. That is, does the natural-draft or mechanical-draft evaluate more favorably? Both the cross-flow and counter-flow types are adaptable to mechanical- or natural-draft. However, the choice is not to decide between the four types but, first, to evaluate between natural-draft and mechanical-draft; then evaluate between counter-flow and cross-flow.

The statement the author makes that the cooling tower is at the mercy of the whims of the weather is a real understatement. Generally, a person can analyze a piece of equipment which he understands and determine what effect certain changes will have on that equipment. This is not so easily done with a cooling tower. Let me present a simple glossary at this point.

Dry-Bulb Temperature.—This is the temperature you read directly from your thermometer on the outside of the kitchen window or the back door, or the temperature which the weatherman gives you.

Wet-Bulb Temperature.—This requires a special piece of equipment called a psychrometer. A simple wet-bulb thermometer is made by placing a cotton wick around the bulb of the thermometer and immersing the end of the wick, not the thermometer, in a container of water. This actually is a way of predicting evaporation. The wet-bulb and dry-bulb temperatures determine the relative humidity, which is the next term.

Relative Humidity.—The relative amount of moisture in the air to that which could be present at saturation for a given temperature. Therefore, the higher the wet-bulb temperature in relation to the dry-bulb temperature, the higher the relative humidity.

Hot Water.—The water to the tower.

Cold Water.—The water from the tower, or water stored in the cold-water well under the tower.

Range.—The difference in temperature between the cold water and hot water, or the degrees removed from the water. In a steam-plant, this is also, of course, the degrees being added to the water by the steam-turbine condenser.

Approach.—The difference in temperature between the cold-water temperature and the wet-bulb temperature.

There are other terms, but these are the most evident. When you consider that each of these is influenced in some way by the other, and you
have at least three of these changing constantly (wet-bulb, dry-bulb, and
relative humidity), then trying to analyze tower performance is not so
easily done. To complicate the problem further, the wind velocity has its
influence; and then, if the purpose of the tower is to cool the water before
returning it to a stream, you have the stream-flow and stream-temperature
influence to consider in the final performance.

In selecting weather conditions to which to design, I believe it is cus-
tomary to select that wet-bulb temperature which is not exceeded over 5
percent of the time during the summer months. However, in some cases,
the yearly average wet-bulb is used; but I believe this can lead to compli-
cations. Often it is not possible to obtain long-term weather conditions at
a given site. The time from site selection to tower design is generally very
short. If data can be obtained for two or three years, this can be compared
with Weather Bureau data of nearby cities and, if comparable, then the
long-term data of the city could, in all probability, be used.

Mr. Waselkow has presented some specification figures for mechanical-
draft towers, as a matter of interest. In the case of the Southwestern Public
Service Company towers, he indicates a cooling range of 14.7° and an
approach of 9.1°. For the Public Service Company of Colorado, he indicates a range of 18.1° and an approach of 20°. For comparison, the TVA
hyperbolic towers at Paradise Steam Plant were designed for the following
conditions:

Hot Water	99.4° F
Cold Water	71.9° F
Wet Bulb	52.2° F

This is a range of 27.5° and an approach of 20.3°. These same towers
have a summertime duty as follows:

Hot Water	122.7° F
Cold Water	95.2° F
Wet Bulb	72.6° F
Range	27.5° F
Approach	22.6° F

I find the account of the usage of sewage effluent as cooling-tower
makeup very interesting. I think this is just the beginning of what we will
see much of in the future. Let us think of pollution in general for a moment. I can visualize a great interchange between the industries, utilities,
and municipalities of the waste products and by-products. What may be

waste for one may be a useful product for the other. For instance, a cement plant must settle out its limestone wastes. Some chemical plants or glass manufacturers must neutralize their acid wastes with lime slurry. Why can't the two get together and settle each other's problems? I can also foresee useful products being manufactured from the ultimate wastes produced from pollution clean-up.

But back to thermal pollution and cooling towers.

Let us now consider the drift losses some more. As the saying goes, there is always some bitter with the sweet. I am not really saying cooling towers are the sweet answer to thermal pollution, but considering that they are the answer in some instances, let's look at some of the problems. As I see it, there are four types of pollution. They are thermal, stream, air, and land. If you reduce any one of the four, it is to the detriment of at least one of the other three. This is where drift from cooling towers comes in: for, you see, that white, steam-like stuff that Mr. Waselkow's friend in Denver complained about is not as pure as it may look. It is mostly fogging or condensation of the evaporated water of the cooling process. But a certain percentage is the drift Mr. Waselkow told us about. Without drift eliminators, the drift can run as high as one percent of the water flow. With drift eliminators, this can be reduced, but of course to some small sacrifice of the efficiency of the towers (McKelvey and Brooke, 1959). The fog will carry no impurities, since it is actually condensate. The drift is very minute droplets of the dirty circulating water, which is carried into the atmosphere by the draft of the tower. So now we have added to the air pollution. How much depends upon quality of makeup, quantity of blowdown, and effectiveness of the drift eliminators? How much air pollution? Well, consider a 2500-megawatt plant with makeup 2 percent at 200 ppm impurities (and that's good makeup) and drift loss 0.2 percent. The solids to the air amount to 4 tons per day. Now, assume that the plant site is located on a fairly large stream which would support the plant with a 10-degree rise, but suppose the standards say 5 degrees. Cooling towers are almost inevitable. Now, look at another type of the pollution mentioned earlier. To resort to cooling towers means the plant must operate at a higher cooling-water temperature during summer months. This causes a lower plant efficiency, which, in turn, requires more coal to be burned, which puts more ash and gases into the air along with that 4 tons of pollution off the cooling tower. So we have eliminated the thermal pollution at the expense of air pollution. You say, "Use better makeup and increase the efficiency of the precipitators." Then you have a solids-waste to pollute the stream or the land.

Incidentally, the pretty white stuff or evaporation out the top of the tower at the 2500-megawatt plant will amount to 30 million gallons of water per day. I'm not sure, but I believe that amount of water would almost meet the daily needs of the city of Nashville. So you can see why, as Mr. Waselkow stated, makeup on a small stream or where water is scarce can also be quite a problem.

Let us see what happens to the drift and evaporates from a mechanical-draft tower and a natural-draft tower. With a mechanical-draft tower, the drift and condensed, evaporated-water or fog is emitted close to the ground and produces local fogging, as described by Mr. Waselkow. In the winter, this can present problems due to freezing. With the hyperbolic natural-draft towers, all drift and fog is emitted over 300 feet into the air. Although some droplets may get back to the ground, the largest percentage is dispersed into the atmosphere so that there are few, if any, localized problems.

In connection with the hyperbolic natural-draft towers, I would like to present to you some of the highlights of the Tennessee Valley Authority Paradise Steam Plant located in Western Kentucky on the Green River.

When Unit 3 is completed in 1969, in addition to adding 1150 mw to the existing plant capacity of 1408 mw, a unique utilization of its condenser-water cooling system will increase the availability of Units 1 and 2 and improve the river-water quality by a part-time reduction of the heat load to the river. Units 1 and 2 were placed in commercial operation in 1963.

It was recognized, in planning the two-unit station, that the plant would not have 100-percent availability because of summertime low flows and high temperatures in the river, the source of the condenser-cooling water; however, its location in the coal-mine area made this site the most attractive because of the exceptionally low cost of fuel and the over-all power-production cost.

Supplemental cooling by cooling towers or cooling pond could not be justified in the original design, due to the short periods of curtailment or shutdown involved.

During the summer months, periods occur when river flow is less than the requirement of 550 cfs for one-unit operation.

Operating guidelines required that the average of the four temperatures at a downstream monitoring station should not exceed 95° F. at any time. When this average temperature was approached, station load was curtailed as required.

In 1965, the decision was made to install Unit 3 at Paradise. This will be an 1150-mw unit with a closed-condenser cooling-water system utilizing natural-draft hyperbolic cooling towers.

It was determined that, by a small overdesign of the cooling towers, they could accept part of the flow from the Unit 1 and Unit 2 condensers during low river-flow periods and thereby increase the availability of Units 1 and 2 to nearly 100 percent, under normal atmospheric conditions.

Bids were accepted on both counterflow- and crossflow-types of towers and towers having either redwood or cement-asbestos heat exchangers. Five bids were received. Competition was very keen, and in the final analysis and evaluation, TVA selected counter-flow cement-asbestos-fill towers. Evaluations were based on cost of additional pumping head of one design over the other, probable difference in maintenance cost of different types of materials, difference in costs of tunnels to and from the inlets and outlets of the towers, and benefit of difference in guaranteed cooled-water temperatures. Although the higher summer flows cause high temperatures of the water for Unit 3 operating alone, the value of the additional availability of Units 1 and 2 greatly outweighs the detriment.

While the cooling-tower bids were being analyzed, a change in criterion of condenser-water discharge temperatures into the Green River was made. The new criterion established by TVA for this river requires a mixed downstream temperature not to exceed 90° F. with no water to be discharged back to the river over 93° F. Although this criterion could be met with two cooling towers, it would have required operating Units 1 and 2 with 405,000-gpm condenser flow, which is considerably lower than design full-flow and would result in higher than normal back-pressure and/or load reduction. This condition is expected to occur, in some years, from May to September. Due to this long period of operation with inadequate water, it was determined that an additional tower identical to the two proposed towers would be desirable. When the contract was awarded, an option for a third tower was obtained. Evaluation of alternate schemes for meeting the new criterion proved the third tower with its additional benefit to Unit 3 to be the most economical solution, and the option was exercised.

The towers are each 320 feet in diameter, at ground level, and 437 feet high. These are presently the largest shells in the world.

Figure 1 shows the system as it will be when completed. An underwater dam and a skimmer wall have been built in the discharge channel for existing Units 1 and 2. A new channel leads from the present channel between the existing discharge structure and skimmer wall and underwater dam to a new intake structure. The skimmer wall and underwater dam allow nor-

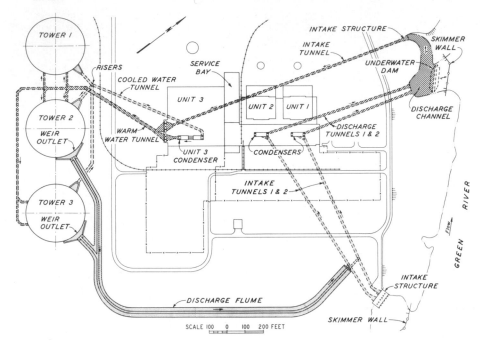

Fig. 1—. Cooling-water system.

mal operation during the cool season, but when flow is being diverted to the cooling towers, the skimmer wall prevents the warmer water from spreading on the surface and escaping to the river. The bottom of the skimmer wall is below the top of the underwater dam, so that a distinct stratified layer between river water of normal temperature and hot condenser-water will exist between the underwater dam and skimmer wall, which is located on the cool side of the dam.

Seven pumps are located adjacent to the Unit 3 powerhouse. Three pumps are for pumping the Unit 3 condenser-circulating water, and four pumps are for Units 1 and 2 supplemental cooling. During normal operation, Units 1 and 2 condensers will be cooled by the present system, which takes water from and returns water to the river. The Unit 3 condenser will be cooled by the water flowing by gravity through an 11-foot-diameter, concrete-lined tunnel to the condenser and then to the circulating-water pump suction. The water will then flow through a 14-foot-diameter, concrete-lined tunnel to the towers.

When the cooling towers are serving all three units, the circulating-water supply pumps at the river intake will be stopped. The four supplemental cooling-water pumps for Units 1 and 2 will take suction from the 1350-

foot-long, 11-foot-diameter, concrete-lined tunnel from the new intake at the present discharge channel of Units 1 and 2. These pumps will discharge into the 14-foot-diameter tunnel along with the Unit 3 flow, and the combined flow of 846,000 gpm will flow to the towers. Since more water will be entering the cooling towers than is flowing back to Unit 3, the water level will be raised in the cold wells and the additional water will spill over 160-foot-long weirs which will be provided in the basin curbs of two of the towers. The overflow will flow, by open channel, to inlet structures which will allow the water to enter the existing tunnels which supply water from the river pumping station to Units 1 and 2 condensers. The water from the towers could have been returned to the Units 1 and 2 river pumping-station intake channel, but then it would have to be pumped back to the condensers. By returning it to the tunnels directly, considerable pumping power is conserved by shutting down the normal supply pumps.

The yearly average cold-water temperature for Unit 3, if operating alone on three towers, would be 66.4° F. The temperature of the re-cooled water for all three units under average summertime conditions will be 90° F., which is only 5 degrees higher than the maximum summertime river temperature. The maximum temperature of the re-cooled water for all three units under maximum summertime conditions will be 100° F.

By providing the present Units 1 and 2 with a summertime closed-condenser cooling-water system, the thermal stream-pollution criterion has been more than adequately met, in that, during the summer months, no water will be returned to the river; therefore, instead of a maximum downstream temperature of 90° F., the temperature will be the natural stream temperature of 85° F.

In connection with the cost of mechanical-draft and natural-draft towers, I believe that, before you can say that hyperbolic towers are too costly, each site has to be evaluated. It seems that somewhere between a 300-megawatt unit and an 800-megawatt unit, there is a break-point that tends to give the hyperbolic wet-type natural-draft an edge over the mechanical-draft type. With the large cooling-water flows, the mechanical-draft towers become quite large and actually require more land area than the natural-draft towers. First costs are in favor of the mechanical-draft towers; however, power costs to operate the fans become very costly and maintenance is much higher for the mechanical-draft towers.

For one specific installation, the first cost of the natural-draft tower was found to be only 1.25 times the first cost of the mechanical-draft towers. When the estimated operating costs were added to the towers, it was found

Fig. 2——. Paradise Steam Plant and cooling towers.

that first costs as high as 1.6 to 1.7 times the first costs of the mechanical-draft towers could be justified (Van Rysselberge, 1959). When you can visualize close to one million dollars in savings through using natural-draft towers, your sense of esthetics tends to change somewhat. The Paradise Towers are shown in Figures 1 and 2.

Gentlemen, as Mr. Waselkow says, "All that is needed is money." I ask, "How much money?" As I stated earlier, the magnitude of the problem is unknown. Senator Edmund S. Muskie (D-Me.), Chairman of the Public Works Pollution Subcommittee, recently said, after hearings held in Florida, "The more we take testimony, the more we are concerned that we need to know more before a change is allowed to take place" (*Electrical World*, 1968).

In Great Britain, where some nuclear plants have been on line for years, biologists have been keeping close tabs on thermal pollution from nuclear as well as conventional stations. After nine years of study, they are sure that the term "thermal pollution" is a misnomer and that there is no evidence of an ecological disaster of any kind which could be attributed to a power station (*Electrical World,* 1968).

At a recent New York state seminar on thermal discharges to watercourses, Howard D. Phillip of Niagara Mohawk Power Corporation stated, "The potential for harm to our environment clearly indicates need for direction and control of thermal discharges. Until much-needed research is completed, it would be unwise to establish firm thermal-discharge standards. The development of interim guidelines would be more appropriate" (*Electrical World,* 1967).

We have some solutions to thermal pollution, but the more stringent the controls, the more it will cost to produce the electrical power that has been one of the greatest contributing factors to the growth of this nation. In the end, "the money" will come from you, the power-consumer.

REFERENCES

Electrical World. 1968. "Britain Doesn't Stew Over Thermal Pollution." 169(Apr. 22):18.

Electrical World. 1968. "It's Not Who, But What on Thermal Pollution." 169(Apr. 29):23.

Electrical World. 1967. "Thermal Pollution Raises Some Government Tempers." 168(Nov. 13):79.

McKelvey, K. K., and Maxey Brooke. 1959. *The Industrial Cooling Tower.* New York: Elsevier Publishing Co.

Maze, Roy W. 1967. "Practical Tips on Cooling-Tower Sizing." *Hydrocarbon Processing* 46(Feb.):2.

Van Rysselberge, J. F. 1959. "Cooling Towers Compared from an Economical Standpoint." *Power Engineering* 201(June 22):3–7.

Chapter **10**   George O. G. Löf and John C. Ward

ECONOMIC CONSIDERATIONS IN
THERMAL DISCHARGE TO STREAMS

APPROXIMATELY 80 percent of all water used by industry is used for cooling purposes. By 1980, the power industry estimates that approximately 10 percent of all the freshwater runoff in the U.S. will be withdrawn for cooling purposes. This increasing intensity of water use for industrial-cooling purposes is imposing progressively severe demands on America's streams. The common practice of withdrawing large volumes of water, pumping it through condensers and heat exchangers where a 10° to 20° F. temperature increase occurs, and discharging the warmed water back to the stream has many consequences, most of which are undesirable. When withdrawals for such purposes were modest, a few decades ago, temperature increases in rivers due to these thermal discharges were usually insignificant. But with the large industrial growth in the United States, particularly during the past decade or so, the effects of increased temperature are becoming important. It has been pointed out by Powell (1938) that the temperature of water to be used for cooling, condensing, and air-conditioning is often of more economic significance than the composition and chemical quality of the water.

In another paper at this symposium (Hawkes), some of the effects of thermal discharges on the ecology of the receiving stream have been outlined. In addition are the detrimental effects, resulting in economic losses, on the downstream users of water for additional cooling and on all other

The work upon which this paper is based was supported by funds provided in part by Resources for the Future, Inc., Washington, D.C., and in part by the United States Department of the Interior as authorized under the Water Resources Research Act of 1964, Public Law 88–379. Permission from the American Water Resources Association to publish extracts and illustrative material previously appearing in the Proceedings of the Second American Water Resources Conference is also acknowledged.

users forced to limit withdrawals because of insufficient river flow. An occasional and limited bonus resulting from thermal discharge may be obtained in northern inland localities, where river navigation may be somewhat improved by reducing winter ice problems (*Chemical and Engineering News*, 1968). Other side benefits may also exist (*Industrial Research*, 1968).

In examining the economic implications of thermal discharge, all the above effects, including sub- and side-effects, should be considered. In this paper, however, attention will be limited to the thermal discharges from steam-electric generating plants, contributing by far the largest total of all heat dissipated into cooling water, and will also be limited to the potential effects of these thermal discharges on subsequent cooling-water users. Although thus dealing with only a portion of the thermal-discharge problem, this selection covers the major source and one of the important consequences to downstream users.

HEAT DISCHARGE QUANTITIES

Thermal discharge can, of course, be avoided by users of cooling water by recirculating the heated water through cooling towers, spray ponds, or natural or artificial ponds where the absorbed heat is dissipated primarily by evaporation. Except where ample land for ponds is available at low cost, cooling towers are usually employed. Warm water from the condensers and heat exchangers trickles down through loose packing in the tower, meeting a rising current of air, usually driven by fans at the bottom or top of the tower. Tall, natural-draft towers may also be used. Evaporation of a small portion of the water results in cooling the balance back to virtually the feed-water temperature. This supply is then recirculated to the plant for further cooling purposes. Small amounts of chemicals may be added to the water to reduce corrosion and fouling of equipment by chemical and biological deposits, and a small percentage of water is continually discharged in order to prevent excessive accumulation of dissolved impurities.

If the water is to be cooled, say 10° F., requiring the dissipation of about 10 BTU per pound of water, the evaporation of approximately 0.01 pound (or one percent) of water will accomplish this cooling. If the air supplied to the cooling tower is much colder than the delivered water, somewhat less evaporation occurs. Thus, if there is no other outflow from the cooling circuit, water intake requirements can be reduced to 1 to 2 percent of those in a once-through system. However, there is usually a certain amount of fine mist or spray carried out of the tower by the air

stream, and, as indicated above, some "blow-down," or liquid water discharge from the circulating stream, must be provided if windage losses are
insufficient. Windage losses from various cooling devices are:

Mechanical draft towers, 0.1 to 0.3%

Atmospheric towers, 0.3 to 1%

Spray ponds, 1 to 5%

Ordinarily, the net-withdrawal requirements range from 2 to 4 percent of
the once-through quantities.

As indicated in Figure 1 (Löf, 1966a), the largest cooling-water use in
steam-electric power plants in 1959 was in the east north-central states,
where, also, the use of recirculation systems is in the smallest proportion.
In the western states, however, where total power generation and water
usage are much lower, recirculation is relatively much more important,
primarily because many of the streams on which power plants are located
do not have sufficient flow for once-through cooling use. Recirculation cooling systems are therefore a necessity. A modern plant of one-half-million-
kilowatt capacity, at an over-all efficiency near 40 percent, will discharge
about 4,000 BTU/(hour) (kilowatt capacity) into cooling water. The condensers will require over 0.6 gallons/(minute) (kilowatt capacity) (with a
typical 12-degree temperature rise) corresponding to nearly 800 cubic feet
of water per second. This is above minimum stream flows in many parts
of the country, indicating the absolute necessity of recirculation-cooling in
those locations. Another alternative, also being employed with increasing
frequency, is a higher temperature rise in the condensers. This permits a
proportionate reduction in cooling-water flow, but at the expense of lower
over-all efficiency.

In addition to the obvious requirement for recirculation cooling by
plants located where there is insufficient stream flow for once-through
operation, recirculation and cooling towers may be the choice, for economic reasons. Even where water flows are adequate, factors may result
in minimum cost being achieved by use of cooling-tower systems. As shown
below, total costs of recirculation cooling, including capital investment in
cooling tower and auxiliaries and all operating expenses, are near one cent
per thousand gallons circulated. This is roughly equivalent to one-half mill
per kilowatt-hour generated. Now, if the cost of supplying cooling water in
a once-through system, including withdrawing (i.e., water purchase), pumping, treating if necessary, exceeds about one cent per thousand gallons,
recirculation cooling would usually be employed. On-site costs would be
minimized by such a decision. This is occasionally the situation, even in

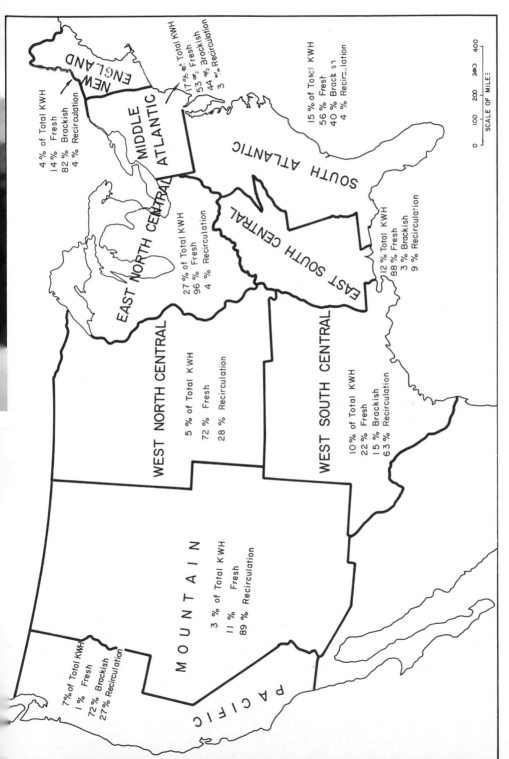

NEW ENGLAND

4 % of Total KWH
14 % Fresh
82 % Brackish
4 % Recirculation

MIDDLE ATLANTIC

17 % of Total KWH
53 % Fresh
44 % Brackish
3 % Recirculation

SOUTH ATLANTIC

15 % of Total KWH
56 % Fresh
40 % Brackish
4 % Recirculation

EAST NORTH CENTRAL

27 % of Total KWH
96 % Fresh
4 % Recirculation

EAST SOUTH CENTRAL

12 % Total KWH
88 % Fresh
3 % Brackish
9 % Recirculation

WEST NORTH CENTRAL

5 % of Total KWH
72 % Fresh
28 % Recirculation

WEST SOUTH CENTRAL

10 % of Total KWH
22 % Fresh
15 % Brackish
63 % Recirculation

MOUNTAIN

3 % of Total KWH
11 % Fresh
89 % Recirculation

PACIFIC

7% of Total KWH
1 % Fresh
72% Brackish
27% Recirculation

SCALE OF MILES

0   100   200   300   400

Fig. 1—. Water use in steam-electric utility plants, 1959.

water-abundant regions, where other considerations have dictated the location of power plants. For example, a mine-mouth power plant may achieve sufficient fuel economies (the largest cost item in power generation), to offset the added cost of recirculation cooling necessitated by stream flows too low for once-through use.

A third reason for recirculation cooling has very recently been encountered. This is in the form of regulations prohibiting unlimited thermal discharge. Until recently, the off-site or downstream benefits of recirculation cooling compared with once-through cooling and resulting thermal discharge have scarcely been recognized. Since the use of recirculation cooling results in heat discharge to the atmosphere, through evaporation, thermal discharge to streams is entirely eliminated. The establishment of standards which directly or indirectly limit the discharge of heated water to streams thus presents an additional incentive for recirculation cooling.

Of growing seriousness, particularly in the last few years, are the off-site damages to downstream users, due to thermal discharge. However, there have been no effective policies for preventing such damages. There has been no incentive for the industrial unit discharging the heated water to incur the cost of abatement. The rapid growth of electric-power generation has, of course, been a significant factor in this problem. Figure 2 illustrates the past increase in electricity generation and the projected totals (Löf, 1966b) to 1980. Also shown are estimates of cooling-water requirements, expected to double in the next 15 years. Estimates of the future growth of recirculation (made prior to the establishment of temperature standards in streams) are shown (Löf, 1966b) in Figure 3. There is little doubt that this recirculation-increase estimate is too low, now that temperature standards are being set and regulations are being established. The recent decision to employ recirculation cooling in a large new nuclear-power plant planned for the upper reaches of the Connecticut River illustrates this trend. It will probably soon be the exception, rather than the rule, for a newly built inland power plant to employ once-through cooling.

Evaporation losses in Figures 2 and 3 were computed from generation and thermal efficiency by assuming 90 percent boiler efficiency and that all heat entering the cooling water results in evaporation at the rate of ¾ pound of water per 1,050 BTU discharged (equivalent to assuming that 75 percent of the heat is ultimately discarded by evaporation, the balance by convection and radiation to the atmosphere).

In 1963, 0.1 percent of the total electricity generated was by means of nuclear reactors, but this figure is expected to increase to 13.3 percent by

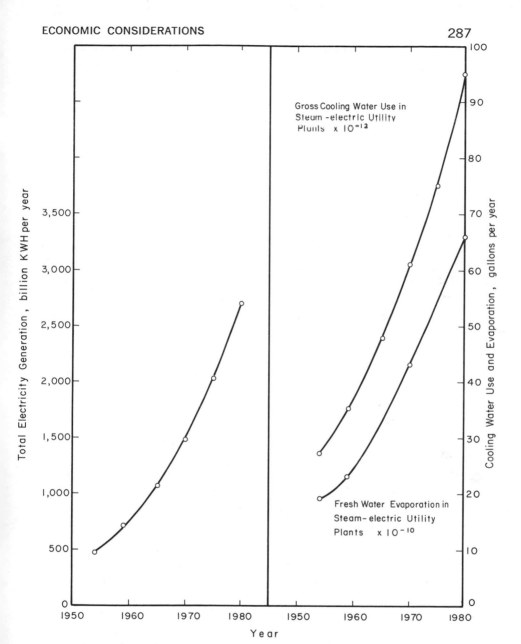

Fig. 2—. Annual electricity generation, cooling-water use, and evaporation.

1980 (Federal Power Commission, 1964). At the present stage of commercial development, nuclear-electric generating plants require as much as 40 percent more cooling water than conventional fossil-fueled electric generating plants. Based on a single 540-megawatt installation (*Environ-*

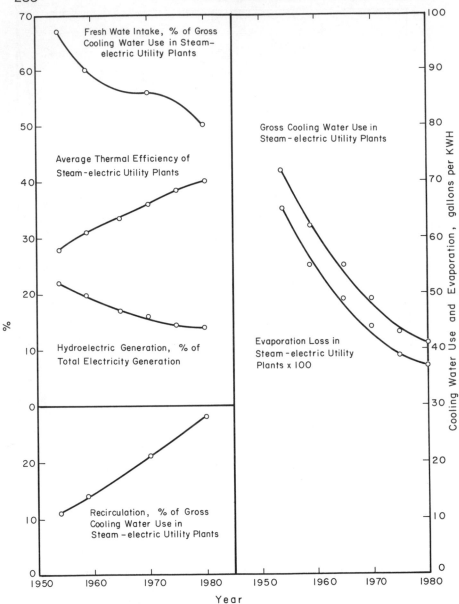

Fig. 3——. Efficiency and water use.

mental Science and Technology, 1968) using large forced-draft cooling towers, the construction cost of the cooling towers was approximately $12 per kilowatt of capacity. Operation and maintenance of these towers is expected to cost $1.67 per year per kilowatt of capacity, resulting in an estimated increase of 0.025¢ per kilowatt hour, which approximates a 1 percent increase in the cost of electricity.

ECONOMICS OF COOLING-WATER RECIRCULATION

It is not the purpose of this paper to detail the technology and accompanying costs of cooling-tower design and operation. References to such works are appended (Cootner and Löf, 1965; Berg, Lane, and Larson, 1964; Ramirez, 1968; Millen, 1968). Rather, the object here is to indicate the approximate range of costs involved in completely preventing thermal discharge, i.e., by cooling-water recirculation through towers, and to show the approximate costs to a downstream power plant of being forced to use cooling water containing thermal discharge from an upstream source.

The cost of cooling-water recirculation is composed of capital costs, primarily of the cooling-tower installation, and operating costs, which are the makeup water, chemicals, and power for operating fans and pumps. The capital costs of the tower are, in turn, dependent on the water flow required, the prevailing wet-bulb temperature of the air, the water temperature change through the tower (equal to the temperature rise through the power-plant condenser), and the temperature of water delivery from the cooling tower to the condenser. On the basis of this information, the capital cost of the cooling-tower installation can be estimated from charts and tables. One of these references (Cootner and Löf, 1965) shows that the cooling of 50,000 gallons per minute from 90° F. to 75° F. at a wet-bulb temperature of 65° F. would require an investment of approximately $640,000. This tower would serve the cooling needs of a power plant of about 100,000-kilowatt capacity. The annual investment expense, C_I, in cents per thousand gallons of water circulated, may then be computed from the equation,

$$C_I = \frac{I(r + 1/t + P)}{5.256N} \qquad \text{(Equation 1)}$$

where

$I =$ *cooling-tower investment per unit capacity, dollars/gpm*
$r =$ *annual cost of capital (interest rate), decimal per year*
$t =$ *cooling-tower service life, years*
$P =$ *annual property-taxation rate, decimal per year*
$N =$ *load factor (fraction of year that cooling tower is used), dimensionless decimal.*

The denominator of Equation 1 is equal to 525,600 minutes/year times N

times 10^{-3} to convert into one-thousand-gallon units times 10^{-2} to convert dollars to cents.

Under typical conditions, the capital cost (C_I) for a forced-draft cooling tower may be about 0.3 cent per thousand gallons circulated. Total capital cost for a forced-draft cooling tower (I) will be about \$8 per gpm times a relative rating factor, K. Ordinarily, $0.4 \leq K \leq 3$. Cooling towers have long service life (t), which is estimated to range from twenty to thirty-three years for large towers.

Relations for costs of operation, C_o, in cents per thousand gallons of water circulated, may be summarized in the following equation (Cootner and Löf, 1965):

$$C_o = 0.001R \left(\frac{C}{C\text{-}1} \right) (0.033Y + 17/C + Wa) + (0.14K + 0.005A)\,p$$

(Equation 2)

where

    R = *cooling range (temperature change of the water passing through the cooling tower)*, °F

    C = *cycles of concentration, dimensionless*

    Y = *alkalinity (as $CaCO_3$) of makeup water, mg/1*

    Wa = *cost of makeup water, cents per thousand gallons*

    K = *relative rating factor of the cooling tower (the relative size of the tower compared with one for the same water flow but operating at a set of "standard" conditions), dimensionless*

    p = *height to which the water must be pumped for flow through the cooling tower, feet*

    A = *cost of electric power, ¢/kwhr*

Assumptions made in the development of Equation 2 are:

1. Evaporation of 0.001 pound of water will cool one pound of water 1° F.
2. The price of sulfuric acid is 4¢ per pound. If the price is 2¢ per pound, the quantity $0.033Y + 17/C$ is replaced by $0.017Y + 13.3/C$.
3. Fan horsepower $= 0.01KL$ horsepower, where $L =$ total water-flow through the cooling tower in gpm. Actual values may range from 0.007 KL to 0.013 KL horsepower. If the tower is natural-draft (no fans), then the term 0.14 K is zero.
4. 0.5 kwhr is consumed in lifting 1,000 gallons of cooling water 100 feet.

5. Fan electric-motor efficiency $= 90\%$.

It is common practice to operate moderate-sized cooling towers with $3 \leq C \leq 4$, but power-plant cooling can be more economical if $8 \leq C \leq 10$. If one chooses values of some of the parameters in Equation 2 as follows, $C = 9$, $y = 150 \ mg/l$, then one obtains

$$C_o = 0.00113R(6.9 + Wa) + (0.14K + 0.005A)p \qquad \text{(Equation 2A)}$$

Values of K (for forced-draft cooling towers) to be used with Equation 1 and in Equations 2 and 2A are given in Table 1. The condenser inlet temperature ($°$ F.) is equal to the sum of the wet-bulb temperature and the approach. The temperature of the water from the condensers before ·cooling is the condenser-inlet temperature plus R. Therefore, the temperature of the hot water is the wet-bulb temperature plus the approach plus R.

TABLE 1. Values of K For Forced-Draft Cooling Towers

| Wet Bulb Tempera- ture, °F | R, °F Cooling Range of | | | | | | | | | | |
| | 10 Approach, °F | | | 15 Approach, °F | | | | 20 Approach, °F | | | |
	5	10	15	5	10	15	20	5	10	15	20
60	2.6	1.4	0.9		1.8	1.3	0.8		2.2	1.5	1.1
65	2.2	1.1	0.8	3.0	1.6	1.1	0.7		1.9	1.3	0.9
70	1.9	1.0	0.7	2.6	1.4	0.9	0.6	3.0	1.6	1.1	0.8
75	1.6	0.8	0.5	2.2	1.2	0.8	0.5	2.5	1.4	0.9	0.7
80	1.4	0.7	0.4	1.8	1.0	0.6	0.4	2.2	1.2	0.8	0.6

Under typical conditions, the operating cost (C_o) may approximate 0.5 cent per thousand gallons. Hence, total costs of recirculation may be about 0.8 cent per thousand gallons. At a common water-temperature increase through the condensers of $12.5°$ F. and a current average power-plant efficiency near 35 percent, about 55 gallons of cooling water have to be circulated through the system per kilowatt-hour generated. Thus, the costs of cooling-tower operation—that is, of water recirculation—may be about 0.4 to 0.5 mills per kilowatt-hour generated, roughly 5 to 7 percent of generation cost or 2 to 3 percent of combined generation and distribution costs.

The use of greater cooling ranges (R), a recent trend, particularly in large plants employing natural-draft cooling towers, has the effect of reducing recirculation cooling costs below these estimates. With a 25-degree cooling range (compared with 12.5 degrees), recirculation rate is halved, tower size is reduced, pumping energy is decreased, and total operating costs of the cooling-tower system per kilowatt-hour generated are reduced to about one-half the previous values, viz., 0.2 to 0.3 mill per kilowatt-hour. Partially offsetting this saving, however, is a reduction in generating capacity and efficiency due to higher condenser temperature, as explained below. The net result of operation at high cooling ranges (up to 30 degrees, recently) in large plants employing cooling towers is a total-generation cost per kilowatt-hour typically about 0.2 to 0.3 mill above the cost in a plant using once-through cooling.

These figures may be interpreted in several ways. If a power plant has chosen to employ recirculation cooling for reasons of on-site economies, i.e., once-through-cooling costs would have been greater, the cost of thermal-pollution abatement may be considered zero. On the other hand, if recirculation cooling has been adopted, even though once-through cooling would have been cheaper, the *difference* in the two costs could logically be considered the cost of preventing thermal discharge. It may be observed that, even in the hypothetical case of zero cost for once-through cooling, the total cost of recirculation cooling, then chargeable to thermal-discharge abatement, would represent no more than 2 to 3 percent of electricity selling price.

Considering now the possible economic loss by a downstream power plant forced to use cooling water warmer than that which would have been naturally available had there been no thermal discharge upstream, several factors are involved. Unless the upstream plant has used the entire river flow in its once-through cooling operation, there will be some dilution of the effluent discharge, with resulting temperature somewhere between the natural river temperature and that of the heated discharge. Secondly, unless the downstream plant is only a short distance away, there will be temperature changes in the river prior to subsequent cooling-water withdrawal. The rate of cooling is, in turn, dependent upon atmospheric conditions, river turbulence, solar radiation, and so on. Given sufficient distance of travel, the river will eventually cool to the same temperature, at the far downstream point, as if there were no thermal discharge. Natural conditions may, of course, cause heating of the river rather than cooling, the artificially added heat being superimposed on natural effects.

The results of a downstream power plant using warmer condenser water than would naturally have been available are a decrease in total electrical generation and a decrease in thermal efficiency, hence, an increase in costs per kilowatt-hour generated. The net additional cost of power-plant operation due to such temperature increase, C_T, in cents per thousand gallons of water used in the condensers, may be determined by use of the following equation (Cootner and Löf, 1965):

$$C_T = C_c \left[\frac{-(1-\eta)\beta y \, \Delta T \, dT_1}{36 n_o (1 - \beta\eta)(1 - e^{-yt}) T_1} \right] \quad \text{(Equation 3)}$$

where

C_c = *the capital cost of the entire power plant, dollars per kilowatt of installed capacity*

η = *theoretical steam-cycle efficiency (Carnot efficiency), decimal*

β = *turbine-generator efficiency, decimal*

y = *fractional decrease in plant load factor per year, years* $^{-1}$

ΔT = *temperature change in the water stream passing through the condenser, ° F*

dT_1 = *temperature increase in the entering cooling water due to thermal discharge upstream, ° F*

n_o = *present load factor of the plant (equals electricity generated divided by rated capacity), decimal*

t = *power-plant life, years*

T_1 = *natural temperature of cooling water, unaffected by thermal discharge upstream, ° F*

Ordinarily, $\Delta T = R$. Also y can be expressed as $\Delta n/\Delta t$. Usually T_1 exhibits a sinusoidal variation during the year (Ward, 1963), and $T_1 + dT_1$ may also exhibit a similar sinusoidal pattern on an annual basis (Curtis, Doyle, and Whetstone, 1964). The sum $T_1 + dT_1$ can be evaluated by the methods given in Appendix I.

The cost given by Equation 3 includes not only the additional fuel required by the power plant, but also the cost of additional capacity to make up for the reduction in output due to slightly smaller electric generation per pound of steam passing through the turbines.

When typical values are substituted in Equation 3, including a temperature rise (ΔT) of 12° F. through the condenser and a 10° F. increase in inlet cooling-water temperature (dT_1) due to thermal discharges from one or more upstream power plants, and if power-plant investment (C_c) is assumed $125 per kw, the additional cost of power generation (C_T) will be about one mill per thousand gallons of cooling water circulated. At a flow rate of about 50 gallons of condenser water per kilowatt-hour generated, the increase in power cost thus becomes 0.05 mill per kilowatt-hour. This cost, unless avoided by cooling-tower use in the downstream power plant, would represent an increase of about one-half of one percent in total generation costs. If, as already found in some locations, river temperatures are artificially raised as much as 20 degrees by thermal discharge, these figures would be doubled.

The net effect of excessive upstream thermal discharge on the costs of an individual downstream power plant are thus seen to be modest, but significant. The results of successive thermal discharge along a river are obviously more serious. If the cost of generating power in each plant is similarly affected along a long river, the aggregate economic loss to the economy is substantial. For example, if, in 1970, ten percent of total power generation were to be affected by an average 10-degree rise in condenser-water-inlet temperature, the cost of generating this 125 billion kilowatt-hours would be increased about $6 million.

It is clear from comparison of the off-site costs of thermal pollution to subsequent users of cooling water for power generation and the on-site costs to a power plant for preventing thermal discharge, that prevention is more expensive. Or in other words, the damages to subsequent cooling-water users are not as great as the costs of preventing the discharges in the first place. In this comparison, it is assumed that the use of recirculation cooling by the upstream power plant would be dictated by regulations, and that internal economics due to avoiding once-through cooling would not be great enough to equal the off-site damages. There are, of course, many other downstream effects of thermal discharges, so this comparison is only of limited significance. Moreover, there would always be some internal economies by use of recirculation cooling, if only in decreasing the usual costs of pumping and piping from the water course.

For the same reason as explained above, there is also no particular incentive for recirculation cooling merely to reduce the temperature of an inlet cooling-water supply. The cost of this operation is greater than the benefit. However, the modest savings in fuel and capital costs achieved by lower water temperature may be an additional incentive, on top of the savings in freshwater-supply costs and, of course, in compliance with thermal-discharge standards where applicable.

Still another incentive, if it should be brought to bear by regulation, would be to avoid charges or penalties for thermal discharge. If power plants were assessed for thermal discharge, such costs would then be effectively added to the cost of once-through cooling. Comparison of this total with the costs of recirculation would then lead to even greater use of recirculation cooling. A charge of ten cents per million BTU discharged, for example, equivalent to about one cent per thousand gallons of cooling water, would nearly always exceed the cost of recirculation cooling. Whether this charge is equitable is not the question here; rather, its use in decision-making is illustrated.

EVAPORATION LOSSES

A further economic consequence of thermal discharge is water loss by evaporation. The heated water returned to the river, with or without dilution from the main flow, results in a higher stream temperature than would naturally be present at that location and at that set of atmospheric conditions. During most of the year, natural river temperatures are below those of the air in contact with the water, but the water temperature is usually fairly close to the wet-bulb temperature of the atmosphere. The lower the relative humidity, the greater is the difference in dry-bulb and wet-bulb temperatures. Normally, wet-bulb temperatures range from a few degrees to as much as 15° or 20° F. below dry-bulb temperatures. Thus, river temperatures may also be below average atmospheric temperatures by nearly this amount. Offsetting this depression, however, is the solar radiation absorbed by the river, tending to raise its temperature and contributing a thermal load.

Whenever the river-surface temperature is above the wet-bulb temperature of the atmosphere, water will evaporate. The rate of evaporation is a function of this temperature difference. If the water temperature should also be above the dry-bulb temperature of the atmosphere, as it often is, in winter, and frequently at night during all seasons, some heat will be transferred by ordinary convection, not requiring evaporation for this portion. A third mechanism of heat loss from the river is by radiation into the atmosphere when the water temperature is above the effective radiation-receiving temperature of the sky. This is often the case where skies are clear and humidities are low. In humid climates during the summer, radiation-heat transfer is comparatively small. Under these conditions, the principal mechanism for restoring the river temperature to its natural level is by evaporation, with resultant cooling. This is effectively the same process that occurs in a cooling tower in a recirculation-cooling system; so, from the over-all water-evaporation standpoint, there is not a large difference in the extent of evaporation on-site with cooling-tower use or off-site in once-through-cooling operations. The off-site evaporation loss is, however, slightly less, due to some of the heat transfer being by radiation, particularly if large river surfaces are exposed and flow is relatively slow. The three mechanisms of heat loss discussed above are expressed quantitatively elsewhere (Hatheway, 1968).

From the foregoing discussion, it can be seen that, although thermal discharge has the effect of diminishing downstream flows, so also does recirculation cooling to avoid thermal discharge. In modern power plants, in which approximately 4500 BTU must be discarded in cooling water per kilowatt-hour generated, in-plant or off-site evaporation is slightly over one-half gallon per kilowatt-hour generated. Figure 3 shows past results and future trends in this ratio and Figure 2 shows the total quantities of water evaporated in the nation's power plants and the freshwater streams affected. Present evaporation from this source aggregates about one billion gallons per day. This is, of course, a relatively small loss in comparison with evaporation losses in agriculture and the natural losses from rivers and lakes. In a few locations, however, where one or more large power plants may be located along the course of a relatively small stream, evaporation loss (here, in cooling towers) may represent a major portion of the normal stream volume.

CONCLUSION

The economic consequences of thermal discharge from steam-electric power plants, the largest users of cooling water, insofar as subsequent cooling uses are concerned, are modest increases in operating costs by the downstream users. These increases are not great enough, however, to justify pre-cooling in the downstream users' plants, unless in-plant economies or thermal-discharge regulations require these plants to employ recirculation cooling. Recirculation cooling is invariably used by plants on small streams where flows are inadequate for once-through cooling. In numerous additional situations, the costs of recirculation cooling, typically approaching one cent per thousand gallons of water, are less than the cost of withdrawing and pumping on a once-through basis. In the large majority of locations, however, generally in water-abundant regions, on-site costs of recirculation cooling in power plants are higher than once-through systems. The costs are also higher than the damages which downstream cooling-water users may suffer through use of warmer cooling water than would otherwise be naturally available. These damages do not include, however, those associated with effects on stream ecology, navigation, or other factors.

Justification for temperature standards in streams cannot rest solely on the economies of alternate methods of cooling. Other considerations appear to be of more significance in establishing limits on heat discharge. But these now seem to be of sufficient importance to dictate wide use of recirculation cooling in future steam-power installations. Even in these cases, the additional on-site costs incurred by the power plant (and passed along, in turn, to the power users), due to recirculation cooling, are only a small percentage of total cost of electricity generation and distribution.

APPENDIX I. EVALUATION OF dT$_1$

Garrison and Elder (1964) derived and experimentally verified the following equation:

$$\frac{T}{T_o} = \frac{\Sigma Q}{T_o U c \rho y_o} x + 1 \qquad \text{(Equation 4)}$$

where

T = *mean water temperature in the control volume (T = T$_o$ at x = 0), °F*

ΣQ = *net rate of heat flow into the control volume per unit of horizontal area, BTU/(ft²) (day)*

x = *distance from the reference station at which T = T$_o$, ft*

U = *mean water velocity, ft/day*

c = *specific heat of water, BTU/(lb) (° F)*

ρ = *density of water, lb/ft³*

y_o = *mean flow depth based on surface width, feet*

In Equation 4, it is obvious that $T = T_1 + dT_1$. If some simplifying assumptions are made, an equation similar to Equation 4 can be easily obtained. The time dt required for the stream to decrease in temperature by an amount dT is

$$dt = \frac{-c\rho V}{q A_H} dT \qquad \text{(Equation 5)}$$

where

dt = *differential time, days*

V = *volume of water underlying the horizontal surface area, A$_H$, of the stream, ft³*

dT = *differential water-temperature change, ° F*

A_H = *horizontal surface area of the stream volume, V, ft²*

q = *the rate at which heat flows from the volume V through the area A$_H$, BTU/(day) (ft²)*

During the time dt, the water in the stream moves a distance $dx = U \, dt$. Therefore

$$dx = -\frac{U c \rho V}{q A_H} dt \qquad \text{(Equation 6)}$$

For relatively short stream distances, q is approximately constant for that small distance. Also the ratio V/A_H can be defined as the effective depth of the stream D, in ft. By analogy with reservoirs (Fair and Geyer, 1954), one might expect D to be roughly one-third of the maximum depth of a stream for a given cross-section. Therefore

$$\frac{T}{T_o} = 1 - \frac{q}{T_o U c_\rho D} x \qquad \text{(Equation 7)}$$

Equations 5 through 7 are a simple rationalization of Equation 4, but are not a derivation. Using the observed values of T and x given by Garrison and Elder for the Holston and Susquehanna Rivers, an empirical relation for the temperature range observed *(63.7$\leq$ T$\leq$ 84.4° F)* appears to be roughly

$$\frac{q}{U c_\rho D} = \frac{T-52}{53,000} \qquad \text{(Equation 8)}$$

a relation which indicates that the temperature decrease may be roughly $\{[(T + 1)/53] - 1\}°$ F per 1,000 feet.

An empirical relationship proposed by Le Bosquet (1946) can be expressed as follows:

$$\frac{T - T_a}{T_o - T_a} = 10^{-0.434} \frac{kwx}{c_\rho U A_c} \qquad \text{(Equation 9)}$$

where

T_a = *air temperature,* ° *F*
k = *heat exchange coefficient, BTU/(ft²) (° F) (day)*
w = *average stream width, feet*
A_c = *cross-sectional area of the stream, ft²*

According to Le Bosquet, the coefficient k varies from 72 to 384 *BTU/ (ft²) (° F) (day)* depending on stream conditions and must be determined from an observed river-temperature profile.

If one assumes that $A_c = wD$ and that $k = q/T_o$, then

$$\frac{T - T_a}{T_o - T_a} = 10^{-0.434} \frac{qx}{T_o c_\rho U D} \qquad \text{(Equation 10)}$$

If the ratio $qx/T_o c_\rho U D$ is appreciably less than one, then Equation 10 can be approximated by

$$\frac{T - T_a}{T_o - T_a} \approx 1 - \frac{qx}{T_o c_\rho U D} \qquad \text{(Equation 11)}$$

Equations 7 and 11 are similar if T_a is near zero ° F or if T is near T_o.

One might assume that the value of k in Equation 9 is given by

$$k = \frac{q}{T - E} \qquad \text{(Equation 12)}$$

where

E = equilibrium water temperature (Edinger and Geyer, 1968),° F
When $T = E$, $q = 0$. Values of k in Equation 12 obtained in a recent
study (Edinger and Geyer, 1968) varied from 144 to 155 BTU/(ft²)(° F)
(day).

APPENDIX II: NOTATION

The following symbols are used in this paper:

A = *height to which water must be pumped for flow through the cooling tower, ft;*

A_c = *cross-sectional area of the stream, ft²;*

A_{II} = *horizontal surface area of volume V, ft²;*

C = *cycles of concentration, dimensionless;*

C_c = *power plant capital cost, \$/kilowatt;*

C_I = *annual investment expense, ¢/thousand gallons;*

C_o = *cost of operation of cooling tower, ¢/thousand gallons;*

C_T = *net additional cost of power plant operation due to thermal discharge upstream, ¢/thousand gallons;*

c = *specific heat of water, BTU/(lb) (° F);*

D = *effective depth of stream, ft;*

E = *equilibrium temperature, ° F;*

I = *cooling-tower investment per unit capacity, \$/gpm;*

K = *relative rating factor of a cooling tower, dimensionless;*

k = *heat exchange coefficient, BTU/(ft²) (° F) (day);*

L = *total water-flow through the cooling tower, gpm;*

N = *fraction of year that cooling tower is used, dimensionless decimal;*

n_0 = *present power-plant load factor, dimensionless decimal;*

P = *annual property taxation rate, decimal per year;*

p = *cost of electric power, ¢/kwhr;*

ΣQ = *net rate of heat flow into the control volume per unit horizontal
area, BTU/(ft²) (day);*

q = *rate at which heat flows from the stream surface, BTU/(day) (ft²);*

R = *cooling range, ° F;*

r = *interest rate, decimal per year;*

T = *mean water temperature in control volume, ° F;*

T_a = *air temperature, ° F;*

T_o = *mean water temperature at $x = 0$, ° F;*

T_1 = *natural stream temperature, ° F;*

dT = *differential water temperature change, ° F;*

dT_1 = *stream temperature increase due to thermal pollution,* ° *F;*

ΔT = *temperature increase in condenser-cooling water,* ° *F;*

t = *service life, years;*

dt = *differential time, days;*

U = *mean water velocity, ft/day;*

V = *volume of water underlying area* A_H, ft^3;

Wa = *cost of make-up water, ¢/thousand gallons;*

w = *average stream width, ft;*

x = *distance from point at which* $T = T_o$, *ft;*

Y = *alkalinity (as* $CaCO_3$) *of make-up water, mg/l;*

y = *rate of decrease in plant load factor n, decimal per year;*

y_0 = *mean flow depth based on surface width, ft;*

β = *turbine-generator efficiency, dimensionless decimal;*

η = *Carnot efficiency, dimensionless decimal;*

ρ = *density of water,* lb/ft^3.

REFERENCES

Berg, B., R. Lane, and T. Larson. 1964. "Water Use and Related Costs with Cooling Towers." *Jour. Am. Water Works Ass'n.* 56 (March):311–329.

Camp, T. R. 1963. *Water and Its Impurities.* New York: Reinhold Publishing Corp.

Chemical and Engineering News. 1968. "Thermal Wastes May Be Aid in Melting Seaway's Ice Jams." (June 17):24.

Cootner, P., and George O. G. Löf. 1965. *Water Demand for Steam-Electric Generation.* Washington: Resources for the Future.

Curtis, L. W., T. J. Doyle, and G. W. Whetstone. 1964. "Discussion of 'Annual Variation of Stream Water Temperature, by J. C. Ward.'" *ASCE, Proc., J. Sanitary Engineering Division* 90(SA4):96.

Edinger, J. E., and J. C. Geyer. 1968. "Analyzing Steam-Electric Power-Plant Discharges." *J. New England Water Works Association* (August):614, 620.

Environmental Science and Technology. 1968. "Nuclear Industry Plans to Control Its Thermal Pollution." (Mar.):165.

Fair, G. M., and J. C. Geyer. 1954. *Water Supply and Waste-Water Disposal.* New York: John Wiley and Sons, Inc.

Federal Power Commission. 1964. *National Power Survey: A Report by the Federal Power Commission.* Washington, D.C.: U.S. Gov't. Printing Office.

Garrison, Jack M., and Rex A. Elder. 1964. "A Verified Rational Approach to the Prediction of Open-Channel Water Temperatures." I.A.H.R., Engineering Laboratory, Division of Water-Control Planning, TVA, Norris, Tennessee.

Hatheway, J. L. 1968. "Equilibrium Surface Water Temperatures." Thesis, Department of Civil Engineering, Colorado State University, Fort Collins, Colorado (June).

Hawkes, H. A. 1968. *Biological and Chemical Effects of Thermal Pollution.*

Industrial Research. 1968. "Heat Pollution—or Enrichment." (July):31.

Le Bosquet, M. 1946. "Cooling-Water Benefits from Increased River Flows." *J. New England Water Works Ass'n.* 60(June):11–116.

Löf, George O. G. 1966. "National Water Requirements for Steam Power-Plant Cooling." *Proc., Second Ann. American Water Resources Conf.* Chicago.

————. 1966. "The Water Demand for Power-Plant Cooling." *Industrial Water Engineering* (Dec.):17.

Millen, C. Keith. "Design and Operation of Cooling Towers."

Ramirez, R. 1968. "Thermal Pollution: Hot Issue for Industry." *Chem. Engineering* (Mar. 25):48.

Ward, J. C. 1963. "Annual Variation of Stream Water Temperature." *ASCE, Proc., J. Sanitary Engineering Division* 89(SA6):1–16.

DISCUSSION/ W. R. Shade and A. F. Smith III

THESE comments are primarily an extension of Professor Ward's paper, and perhaps a repetition of much that has been said. For those of us who make our living in the power industry, the problems of air pollution and water pollution are of paramount importance. Power industry growth, coupled with growing restrictions on air and water pollution, presents major engineering challenges for the future.

We mention air pollution at this water pollution conference for two reasons: (1) Many of the recent decisions to install nuclear generating facilities were strongly influenced by the desire to avoid air-pollution problems of fossil-fired plants; however, the present generation of nuclear units reject 40 percent more heat than a conventional fossil-fired unit, thus aggravating the water-pollution problem. (2) Rejection of heat to the atmosphere to solve water-pollution problems may soon be judged an air pollutant.

Most, if not all states, spurred by pressure from the federal government, have, in effect or under consideration, restrictions on the quantities of heat which may be discharged to natural bodies of water. These regulations have all been phrased in terms of temperature and, for all practical purposes, are sufficiently stringent to preclude the direct rejection of heat to all natural bodies of water except the ocean. New methods for rejecting heat need to be found and these in turn affect the economics of power generation and it is to this question that this discussion is directed.

Every engineering decision involves an evaluation of costs, and costs are directly related to physical considerations. Before discussing the economics of the problem, we would like to discuss some of these basic physical factors. Much has been written on the subject of thermal pollution, and some rather salient and basic facts seem to have been missed; in fact, the excessive concern about temperature limits has tended to obscure some important principles.

302

BASIC PRINCIPLES

1. All Heat Engines Reject Heat to a Heat Sink.—The second law of thermodynamics requires that all heat engines reject heat to a heat sink or reservoir. It should be noted that heat engines include thermal-power plants, gasoline engines, jet engines, and any other engine dependent on burning fuel. In general, the heat rejection is in the order of half to two thirds of the fuel consumed.

2. The Atmosphere Is Our Ultimate Heat Sink.—Many engines, such as automobile and jet engines, reject heat directly to the atmosphere. In thermal-power plants, heat is rejected to various bodies of water, which, in turn, reject this heat to the atmosphere by various heat-transfer mechanisms. We hasten to add that our use of the atmosphere as the ultimate heat sink may not be scientifically correct. The scientist may well regard outer space as the ultimate heat sink and radiation to a "black body" as the ultimate heat-transfer mechanism.

3. Temperature Differential Required for Heat Transfer.—It is to be noted that our problem is heat rejection, and it is a thermodynamic fact that heat transfer requires temperature differentials. All of the regulations on thermal pollution are limiting in terms of temperature, rather than heat, and there is a subtle but important difference. It is our considered opinion that efforts to reduce cooling-water discharge temperatures by designing for low cooling-water-temperature rises and by various dilution schemes only slow down the process of ultimately rejecting the heat to the atmosphere. We do recognize, however, that such schemes may make it possible in certain instances to live with present thermal regulations. Caution is recommended on new plant designs because the regulations may become more restrictive before the plans are approved.

4. Importance of Evaporation.—Heat is transferred from a body of water to the atmosphere by conduction, convection, radiation, and the combination called evaporation. The relative importance of these depends on atmospheric conditions and the solar heat load. Our calculations indicate that, on the *annual average basis,* approximately 50 percent of the heat imposed on a natural body of water is rejected by evaporation. During winter months, when the air temperature is low and the sun is in the southern hemisphere, two-thirds of the heat is rejected by convection and radiation and one-third by evaporation. On an average summer day, when all the conditions are worst, evaporation accounts for more than two thirds of the heat transfer.

5. Thermodynamic Problems of Direct Air Cooling.—If it is accepted that the atmosphere is the ultimate heat sink, why not reject the heat directly to the atmosphere in a manner to avoid actual evaporation and its associated problems? A cursory investigation indicates that such a scheme would be very costly to a large modern power plant, for the following thermodynamic reasons: (1) An evaporative cooler is limited by wet-bulb temperatures; the direct-air cooler is limited by dry-bulb temperatures. (2) Transfer of heat to air by evaporation absorbs 1000 BTU/ pound of water evaporated. To transfer 1000 BTU to air from a heat exchanges requires heating 260 pounds of air (3460 cuft) 16° F. (3) A steam-to-water condenser has a heat-transfer coefficient of approximately 600 BTU/sqft/° F. A water-to-air cooler has a heat-transfer coefficient of approximately 25 BTU/sqft/° F. It should be noted that all of the above factors are in the wrong direction to provide for economical design of dry cooling towers.

COOLING SYSTEMS : COMPARATIVE COSTS

Professor Ward has discussed the "Economic Considerations in Thermal Discharge to Streams" primarily from the standpoint of costs involved in completely preventing thermal discharge by cooling a captive supply of water in a closed system by means of cooling towers, as compared to the costs to a downstream power plant being forced to use once-through thermal discharge from the upstream plant. The authors correctly conclude that the cure by the upstream user is costlier than the damage to the downstream user. The authors also conclude that, in the majority of cases, the on-site costs of recirculation cooling in power plants is higher than once-through systems. This is also correct, *but* the power industry now faces legislation which essentially prohibits once-through cooling and the question that needs answering is: How great are the additional costs of recirculation cooling?

We have compiled a set of comparative cooling-system costs for six different methods of heat rejection for a modern power plant consisting of two 900-mw units with fossil-fired boilers. We have selected this plant because we have actual, present-day costs of equipment for a plant of this size with natural-draft cooling towers. We emphasize that this compilation of costs can be used only for *general comparison*. The costs for any specific project are greatly influenced by local geographic and topographic conditions which may make any one method superior to others. The six heat rejection methods are as follows:

1. Run-of-river cooling
2. Bay/lake cooling
3. Natural-draft cooling towers
4. Natural-draft cooling towers and makeup reservoir
5. Cooling pond
6. Dry-cooling towers

We agree with Professor Ward that once-through cooling is cheaper than recirculation cooling, and we have taken the run-of-river system as the base against which all other schemes are compared. Following is a description of each scheme and an indication as to how it compares with the base scheme.

Run-of-River Cooling System

This scheme assumes an ideal site and contemplates taking water from a river, passing it through the condenser with a 17° F. rise and returning to the river with no recirculation. Our study indicates that the condenser-circulating water pumps, screen house, and circulating-water piping installed will cost approximately $9 million, or $5 per kilowatt of capacity. The yearly average condenser back pressure will be 1.2 inches of mercury, with a 55° F. inlet circulating-water temperature.

Bay-Lake Cooling System

This scheme contemplates taking water from a bay or lake and returning it at a great enough distance from the intake to prevent recirculation. In general, the equipment and structures are identical to the run-of-river scheme except for a long intake or discharge conduit. This cooling system we have estimated to cost approximately $11 million, or $6 per kilowatt of capacity, or $1 per kilowatt over the base. The yearly average back pressure will again be 1.2 inches of mercury.

Natural-Draft Cooling Tower—Run-of-River Makeup

This scheme incorporates natural-draft cooling towers and makeup to the towers from a river or lake. The average annual circulating water from the cooling tower is assumed to be 70° F. with a 28° F. rise through a dual-pressure condenser. In this arrangement, the cooling towers, makeup facilities, and larger condenser are all additional items of cost. The circulating-water pumps, pump house, and piping are all lower in cost than for the once-through system. We estimate the total cost of the cooling system for this arrangement to be approximately $13.5 million, or $7.50

per kilowatt of capacity or $2.50 per kilowatt over the base. The condenser back pressure will be 1.5 inches of mercury in the first section and 2.16 inches of mercury in the second section.

Natural-Draft Cooling Towers: Reservoir Makeup

In many cases, particularly at mine-mouth plant locations, there is not sufficient run-of-river supply to provide cooling-tower makeup during low-water conditions. In such cases, it is necessary to provide a storage reservoir at an appreciable increase in total cost. In a typical case, the land and dam costs for providing a reservoir of this type amounts to an additional cost of $6 million. Therefore, for this case, the cost of the cooling-water system amounts to $11 per kilowatt, or $6 per kilowatt over the base. The condenser back pressure is obviously the same as in the previous case.

Cooling-Pond System

A recirculation system which rejects heat to a captive body of water is a practical solution when the geographic and topographic conditions are favorable. The costs of such a scheme are difficult to estimate because of variable land and dam costs. For the purposes of this comparison, we have assumed a water-surface area of one acre per megawatt, a land requirement for the pond at 20 percent greater than the water surface and a cost of $1000 per acre of land. If this cost per acre seems high, it is because there are, invariably, development costs on such projects, such as: roads to be relocated, bridges to be built, structures to be razed, and wells to be plugged. Past calculations we have made for cooling-pond equilibrium temperatures with a heat load equivalent to one megawatt per acre indicate an average yearly inlet circulating-water temperature of 65° F. in Pennsylvania. Our estimates indicate the condenser- and cooling-water costs for this arrangement to be approximately $18.5 million, or slightly over $10 per kilowatt, or $5 per kilowatt over the base.

With an annual circulating-water inlet temperature of 65° F. and 32° F. rise, the condenser back pressure will be 1.4 inches of mercury in the first pass and 2.2 inches of mercury in the second pass.

Dry-Cooling Towers

Direct rejection of heat to the atmosphere has excited the imagination of power-generating people, particularly in areas where little or no water is available. This has been done successfully on stations in Europe with air-cooled condensers and also with the system at the 120-megawatt Rugeley Station in England.

In general, it has been concluded that using an air-cooled condenser on a unit larger than 150-megawatt size is impractical because of the physical problem of getting the large volume of exhaust steam to the condensing surface. The Rugeley scheme, which uses a jet condenser for the turbine with the spray water cooled by a water-to-air exchanger, seems practical. Whether this scheme is economically feasible is the question. U.S. cooling-tower suppliers have been studying this problem and, to date, the best information we have received is that a dry-cooling tower may cost $20 per kilowatt. A "blue sky" estimate for the condensing and cooling system for our 1800-megawatt plant could easily reach $40 million, or $22 per kilowatt, or $17 per kilowatt over the base. There are many unsolved technical problems, including methods of coping with the wide seasonal variations in dry-bulb temperature, turbine and cycle design for high-back-pressure operation, jet-condenser design in large sizes, physical arrangement of an enormous amount of air-cooled heat-exchange surface with a means of forcing or inducing astronomical amounts of air across this surface.

CONCLUSIONS

From Professor Ward and Mr. Löf's paper and from our own experience, we have arrived at the following conclusions:

1. Run-of-river and bay/lake systems for rejecting power-plant thermal discharge have a very limited future because of government restrictions.
2. Cooling by means of closed systems such as cooling towers and captive cooling ponds is not prohibitive, either in initial capital costs or operating costs, but these costs are not so low they can be ignored.
3. Low seasonal water-flows need not restrict power-plant locations, because reservoirs adequate for cooling ponds or makeup storage for cooling-tower operation can be built in many parts of the U.S.
4. Sites for reservoirs for storage of water for cooling-water makeup purposes and sites for cooling ponds need to be preserved for future development.
5. Dry-cooling towers do not seem to be economically feasible at this time; however, experimentation and development must continue for the day when cooling-tower plumes are judged on air pollutant, and for areas where water is not available.

DISCUSSION FROM THE FLOOR

John W. Lebourveau: From your comparison of natural-draft with forced-draft cooling towers, was your choice of natural-draft towers influenced by the carrying charges? How much were the charges?

Gordon L. Ford: I didn't personally make this comparison, and I really can't get into this question intelligently. My responsibility is in the design of the system.

Vern W. Tenney: Mr. Waselkow, I would like to know the rule-of-thumb figure on percent of blow-down from cooling towers that you talked to us about, and whether or not, in the recirculating water, you used corrosion inhibitors such as chromates, which may, in turn, become a water-pollution problem via the blow-down route?

Charles Waselkow: On the system, I am applying water of a hardness, say, of 130 parts hardness and total dissolved solids of some 350 PPM. In other words, a three-to-one ratio. We blow down 240 GPM on a 100,000-kilowatt unit. The total makeup, including the blow-down, is 1400 GPM. These are actual measurements from our operating meters.

  Your other question was concerning chromates. We do not add any chromates to the type of water we have. Maybe some other people have different water conditions. All we do is add acid to the circulating water, to restore the pH, and then we treat with chlorine to kill the algae. Those are the only two chemicals that we use.

H. R. Drew: There have been some conflicting statements about the relative water consumption in cooling towers, lakes, and rivers. Our experiments at Lake Colorado City, Texas, and other studies by the U.S. Geological Survey have shown that, when heat is added to a reservoir, approximately 50 percent is dissipated by evaporation, the balance being dissipated by other means which do not result in the consumption of

308

water. Probably the same mechanism applies in the case of rivers used for cooling, although this has not been as thoroughly studied. Now, your statement earlier today was to the effect that the water consumed in lakes was 50 per cent greater than that consumed in cooling towers, whereas our studies and experience indicate exactly the opposite: that cooling towers, because they cool almost entirely by evaporating water, consume at least twice as much water as lakes or rivers. I think you were referring to lakes that are built solely for cooling purposes, which is not the typical case. Usually, artificial reservoirs are built for other purposes in addition to cooling, and in the case of natural reservoirs or rivers, that evaporation goes on, anyway. Only the water lost due to the addition of heat can be considered to have been caused by power-plant cooling, and this is much less than when cooling towers are used.

Waselkow: You are quite right in your remarks. Remember, I emphasized the fact that in Colorado and in Amarillo, Texas, the conditions were the same. I am speaking of a lake such as Valmont Lake, which was entirely made for cooling-water purposes for our plant, and the water losses are in addition to that due to the plant operation. There is a certain amount of evaporation by virtue of exposure to the climatic conditions which has to be made up and charged to this installation. We have found that we are actually adding this 50 percent more than we would for an equal cooling-tower capacity. I am speaking of this combination due to the heat input and the natural evaporation.

John Foerster: Mr. Shade and Mr. Ward, how do you equate the following example in your total-systems approach to total-cost analysis for straight-through plant-cooling operation? You have several power plants located along a river course; there are proper mixing zones established; there is adequate cooling water available, but the effect of the thermal effluent on micro-organisms such as algae and protozoa is to shock them into synchronous growth. This shocking eventually produces a bloom, which your plant, which is downstream from one or more similar facilities, draws in. The result is the clogging of intake screens, condensers, etc.

John C. Ward: Of course, we carefully limited the scope of the paper so as not to cover the possibility, because of the difficulties it does open up. Frankly, it would be a highly geographic situation. However, as I indicated in the paper, it might be cheaper to go to recirculation cooling if you have a dirty stream, and with this sort of condition taking place, it may actually be cheaper in operating cost. In fact, I was listening to

Mr. Shade's figures here, and I remember one figure that I used in the paper to illustrate costs, and to the best of my memory it was a 100,000-kilowatt plant, and the stated cost of cooling tower, et cetera, would be $640,000 for a given set of wet-bulb and so-forth conditions, which would be about $6.4 per kilowatt of installed capacity, which is not too much more than the $5-per-kilowatt of installed capacity for once-through cooling that Mr. Shade gave. But, of course, I don't know. It would be a pretty difficult project, economically, to evaluate the concentration of algae and the direct cost. It's a lot easier, thermodynamically, to evaluate increased temperature, but it could well be that the once-through costs are substantially more, especially if you have all these big models.

W. R. Shade: I think we in the power industry have been dealing with dirty streams for many years. We chlorinate and screen. We have recently adopted condenser-cleaning practices which came from Germany, which amounts to a continuous pumping of rubber balls through the condenser. There is no question that debris of any kind, whether it be floating debris or shellfish or algae, is a problem, and how much it costs is a very individual situation, and it may well be that we are going to find out on some of the recirculation schemes that we have some tremendous new problems. In the particular stations that we are dealing with in the western part of Pennsylvania, our big problem is iron oxide. All of the streams that we are on, through the coal regions, and because of the sulphur in the coal, are acid. The acid water dissolves iron, which fouls heat exchangers, and it is going to be a major problem in the future to keep these plants clean.

Emory H. Hall: My question is to Mr. Shade, since he gave the comparison of costs. Do I understand you correctly, that these are the installed, on-site costs? Do they include interest during construction, builders'-risk insurance, general overhead, architect and engineering fees, etc., or not?

Shade: These are installed costs. We have not included the so-called "below the line costs," interest during construction, etc. These items, of course, modify all of these costs by a ratio of their magnitude. Ours are installed costs only, and for this reason may look a little low. They are obviously lower than other figures that have been quoted.

Edward Silberman: I have a somewhat broader economic question. Mr. Shade mentioned that if we stick to the earth, the atmosphere is the ultimate heat sink. I am not sure that is true. I think the oceans are a good reservoir. We have also pointed out that nuclear power was coming to the forefront. It will be a large percentage of production. I wonder if anyone has made an economic study of the feasibility of generating all power on the oceans, using once-through flow, and using a distribution network across the country so that the interior power companies become mainly distributing companies. We would have a few large producing companies on the oceans and probably some pump-storage installations in the interior. It would be worthwhile to consider the economics of this as an alternative to building power plants all over the country.

Shade: Nobody that I know of has made a feasibility study of it. We have thought a lot about it because it always comes up, that the ocean is a tremendous, available heat sink. But consider the problem of getting to the ocean. This is easier said than done. You look at our East Coast, for instance, and you will find that we have an inland waterway. Now, what do we mean by the ocean, and are we going to destroy all of the summer resorts that stretch from Maine to Florida? The costs of this are going to be terrific. I can give you an example: the Oyster Creek Plant in New Jersey, which is a nuclear plant, has suddenly found that it has to go to the ocean for its water, or at least return the discharge to the ocean. They were far down the line in design, expecting to use water from near the shore and reject it near the shore; but with thermal-pollution restrictions staring them in the face, they decided that they were going to have to return the water to the ocean, and this is going to be a multi-million-dollar expenditure for them. Now, the problem of building a power plant in the ocean is one that we haven't faced, but it may well be that we are going to have oil rig structures out beyond the three-mile limit, and then transmit power in. This brings us to another question, of course, and that is power transmission, and I am no authority on this. You may have read newspaper accounts of the attempts—which they had to abandon because of transmission costs—of the power company in New York City to bring power from Canada; but also, because they were bringing high-voltage transmission towers across such areas of the country as Westchester County, New York, which is the richest residential area in the world. Transmission costs are basically the economic problem here, I believe. It is feasible, probably, but the transmission costs are a tremendous factor.

Waselkow: With reference to the possible connection between the East and

West Coasts, I would like to remind you that you may now be sitting under the lights lit by electricity generated in Denver! We are already interconnected through Yellowtail in Montana. We have in this country some 46 million kilowatts on the West Coast and some 120 million kilowatts on the East Coast. We are already operating on the same frequency —same system, you might say, up to a certain point. It's quite critical when you have 46 million kilowatts on one side of a cross-connection and 120 on the other to remain interconnected. Eventually, when the systems are interconnected more than they are now, we can expect a good deal more stability.

Norman H. Brooks: I wonder if Mr. Ford could tell us what the installed cost of the cooling towers at Paradise are; and has TVA published an analysis of what went into making the decision to install cooling towers at Paradise?

Ford: I will answer your questions in reverse order. No, there has not been any publication of this study. No. 2, I think the final cost of the three towers is somewhere in the order of $9 million to $10 million.

Chapter **11**   Frank L. Parker and Peter A. Krenkel

SUMMARY AND STATUS OF THE ART

WE have now been through two symposia on thermal-pollution problems. The first, held in Portland in June, dealt with the biological effects; and the second, which we are concluding today, has concerned itself with the engineering and economic aspects.

In addition to the information that has been gathered from these two conferences, we here at Vanderbilt have also been conducting a critique and the preparation of a "Status of the Art" paper for the Federal Water Pollution Control Administration. Though some of the material may be a bit redundant, I think it is useful at this point to summarize where we stand on thermal-pollution problems and to try to get some insight into what the problems, directions, and possible solutions are going to be. The paper will be followed, of course, by Tichenor's and Cawley's paper on "Research Needs for Thermal Pollution Control." Therefore, this is a summary of where we stand and a preview to lay the foundation for their paper on research needs.

The first question we must ask ourselves is: How serious is the thermal-pollution problem? What is the magnitude of the problem? The greatest single source of heat addition to our waterways is from electric-generating plants. Analysis of steam-electric cooling discharges for 1965 indicates an average 13° F. rise in water temperatures after passing through the condensers. The amount of water withdrawn for this purpose is approximately 42 trillion gallons per year, which is roughly 10 percent of the total flow of waters in U.S. rivers and streams per year. Industrial cooling-water discharges are approximately 20 percent of the total water used for cooling waters, or we might say, approximately 10 trillion gallons per year. The magnitude of this volume is, however, fraught with a great deal of uncertainty because of the great difficulty in evaluating and obtaining definitive information on industrial cooling-water uses. The plants differ widely among themselves, even within the same industry. The temperature of the discharge waters is less critical, and consequently, the information is not as satisfactory as that available from the steam-electric generating plants. We can, however, conclude that this source will not be as concentrated nor as troublesome as are discharges from electric-generating plants. Their relative impact can be assessed by the Federal Water Pollution Control Administration's estimated costs of total cash outlays from 1969 through 1973 for each industry; for cooling steam-generating stations' discharges, approximately $2 billion; and for cooling manufacturing discharges, $700 million.

The electric-generating industry has been doubling in magnitude every 10 years. This rate of increase is projected to continue at least until the year 2000. It might be instructive to glance at Figure 1, which indicates the rate of increase of electricity generation over the last century and the growth of the Gross National Product. The growth in the production of two basic industrial materials, such as steel and sulfuric acid, and the population growth of the U.S. are also shown. We can see from Figure 1 that the rate of increase of electrical use is startling, but that our Gross National Product follows almost the same course (in constant dollars), and has practically doubled every 10 years in the last two decades. The rate of growth in electric usage is due not only to the increase in population, but also to the use of electricity per family unit, which increases as we go to electric space-heating, wider use of air conditioning, and a plethora of electrical appliances of dubious utility.

Another factor which intensifies the thermal-pollution problem is that more than 50 percent of the generating capacity ordered in the last three years has been nuclear, primarily light-water reactors. The thermal efficiencies of the light-water reactors contracted for recently have been on the order of 34 percent in comparison to the 40 percent for equivalent fossil-fueled plants. As long as light-water reactors predominate, and it seems likely that they will for a number of years, we can expect the thermal efficiencies of the nuclear-power plants to be below that of equivalent fossil-fuel electric-generating plants. The reason for this is that the AEC has dropped many of the projects for advanced converters and for superheat and reheat of the light-water fuel reactors, because of metallurgical problems, and seems to be going primarily for the fast-breeder reactors as its next major step.

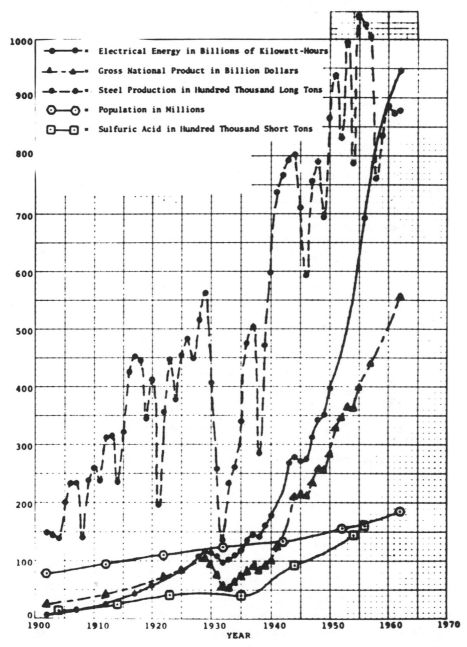

Fig. 1—. Growth in the use of electric energy and selected products, 1900–1970.
(Photograph by Vanderbilt Medical Center Dept. of Medical Illustration)

TABLE 1. Nuclear Power Costs

Year	Atomic Unit	Size (megawatts)	Cost of Station (per kilowatt)	Total Production cost (mills per kilowatt-hour)	Competitive point (cents per thousand B.T.U.)
1964	Post-Oyster Creek	605.0	$139	4.48	26.3
1965	Post-Dresden 2	800.0	123	4.42	24.0
1966	Browns Ferry 1, 2	1,100.0	115	3.78	16.0
1967	Browns Fer.y 3	1,100.0	132	4.04	19.0
1967	Bridgman	1,100.0	139.50	4.30	21.95
1967	Surry	815.5	152.55	4.55	24.83

SOURCE: After Phillip Sporn, "Nuclear-Power Economics: Analysis and Comments, 1967." In **Nuclear-Power Economics, 1962–1967: Report of the Joint Committee on Atomic Energy,** U.S. Congress, 90th Cong., 2d sess., 1967, p. 20.

However, the preponderance of nuclear commitments may be on the wane, as the rise in costs of nuclear-power plants has been considerably above that for fossil-fueled plants, since the low of the Browns Ferry first and second units as shown in Table 1. At that time, nuclear-power plants were competitive with fossil fuels at 16 cents per thousand BTUs. The most recent large nuclear-power plants were competitive with fossil fuels at 24.83 cents per thousand BTUs, which shows the rate at which nuclear-power costs have increased, relative to fossil-fueled plants.

We can also expect that the thermal problems associated with cooling-water discharges from steam-electric generating plants will even further exceed the problems from industrial sources, because it appears that industrial plants have, by and large, already reached the maximum economically efficient size. Decentralization seems to be more important now, whereas, in steam-electric generating plants, we may note that, in the last 15 years, the maximum size of the generating units has risen from 208 mw to 1100 mw. In addition, the increase in size of individual units has made a great difference in the magnitude of the *intensity* of the thermal-pollution problem. Although the total amount of heat disposed of to the environment may not increase when a single 1100-mw plant is built rather than two 550-mw plants at separated sites, the impact is far greater because the amount of dilution immediately available is greatly decreased. The units being installed at Brown's Ferry will be 1100 megawatts and, consequently, even if the unit rate of release of heat were the same, the increase in size of units means that a greater amount of heat is being discharged to the same size stream than would have occurred 10 to 12 years ago with the release from only a 200-mw unit.

It is also true, of course, that the efficiency of these units has increased over the years. The average net-heat rate dropped from 25,000 BTU per kilowatt hour in 1925 to 10,453 BTU per kilowatt hour in 1965; so we have seen a net-heat rate decrease at a 2.8 percent annual rate over the last 40 years, whereas total power generation has increased at a 7.2 percent annual rate and the unit sizes have increased at an 18 percent annual rate over the last 15 years. At present, we seem to be approaching the upper limits of efficiency in fossil-fuel steam-generation at a heat rate of about 8500 BTUs/kwhr, which is an over-all efficiency of about 40 percent. This is the most economic efficient rate. It is possible to arrive at a higher efficiency than this, but the total costs are greater and, therefore, this is most likely the best plant-heat rate that can be expected in the near future. If we compare this 40 percent efficiency with the Carnot cycle, using again the most economic temperature of 1050° F. and the average temperature of surface waters in coterminous U.S. of 55° F., we then arrive at a Carnot cycle efficiency of 66 percent. Therefore, our 40 percent efficiency for the most economically efficient plant is actually 61 percent of the maximum possible efficiency. It cannot be expected that in this cycle the efficiencies can be driven much higher.

The best efficiencies for present-day light-water reactors is about 34 percent, which means that the over-all amount of heat discharged to receiving waters from nuclear-fuel power-generating stations is greater than from fossil-fuel power-generating stations, but the problem itself is not substantially changed. We can, however, look forward to improved efficiencies in the high-temperature gas reactors and in the breeder reactors when they do come on line.

From this fairly extensive discussion of the future growth of the thermal-pollution problem and the sources of this problem, we can see that there is not much hope with the present technology and foreseeable technology to decrease to any substantial degree the total amount of heat rejected to streams from these central steam-generation plants. The interdepartmental energy study under the direction of Ali Cambel stated, "Further cost reductions in conventional central power stations will be difficult to attain in view of the high internal efficiencies and the economies of plant size already realized."

We might also glance for a moment at some advanced technologies for

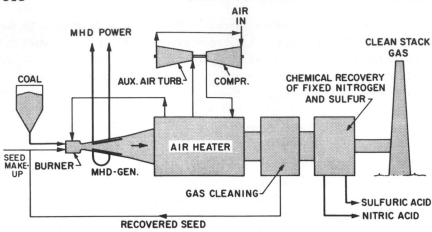

Fig. 2——. Magnetohydrodynamic power-generation system.

(Photograph by Vanderbilt Medical Center Dept. of Medical Illustration)

central power stations. First, at fuel cells, and we can state rather categorically that they are unlikely to compete with large stationary power plants in an industrial nation. Second, at magnetohydrodynamic power generation, as shown in Figure 2. The potentialities for this system are exceedingly high. Although many engineering problems remain to be solved, the Soviet Union is building a 75-mw pilot plant in Moscow, and a 20-mw unit is installed at the Tullahoma (Tennessee) Arnold Engineering Development Center for hypersonic wind-tunnel use. The MHD systems with single cycles can achieve thermal efficiencies of 50 to 55 percent, which possibly could be boosted to 60 to 70 percent by using a binary cycle. If a binary cycle using a gas turbine is utilized, the thermal-pollution problem is avoided entirely. However, even the most ardent advocates of MHD agree that central-station use is at least 10 years away.

Another possibility for central-station generation of electricity is by controlled thermonuclear fusion. Fusion is extremely attractive, as it would provide a virtually unlimited source of energy. However, before a satisfactory system is ready, a great deal of work in plasma physics and engineering is required. In particular, a high-density plasma must be confined and heated to thermonuclear temperature. The engineering development of the vacuum wall, blanket, and chemical processing plant must then be carried out. Even the most optimistic fusion partisans believe that it will not be until the 1980s before fusion power is at the state of fission power today.

The role of thermal electricity and thermyonics in central-station research seems to be limited.

Therefore, we may conclude that the amount of waste heat to be released will continue to grow, and the hopes of any great reduction by increased efficiency will be negated by the over-all rise in the utilization of electrical power.

The possibilities of dilution must be considered next. The average flow of water in U.S. rivers and streams is about 440 trillion gallons per year. It does not seem likely that there is any great possibility of substantially increasing this amount. It would be possible, through the use of low-flow augmentation techniques, etc., to even out the flow during the course of the year, but this still would not solve the problem. Consequently, one may conclude that the availability of an increased amount of water for dilution and lowering of the temperature of the heated discharges is not a very fruitful source of research. It is extremely important, however, that we know the fate of the excess heat discharged to the waterways. The two methods used, to date, to determine the fate of the heat have been the energy-budget approach and the physical model studies in cases (practically all of them) where the boundary conditions were extremely complicated.

The energy-budget method predicts the net change in temperature in a body of water by adding the heat entering the body of water with the incoming water to the heat entering from short-wave solar radiation, long-wave atmospheric radiation, and the heat lost or gained by conduction from the water to the air. From this increased heat load, reflected solar radiation, reflected atmospheric radiation, evaporative heat losses, long-wave back radiation, and heat in the outflowing waters are subtracted. Depending upon time of day, year, latitude, temperature, and wind conditions, different mechanisms may dominate. Usually, net solar and atmospheric radiation, evaporative losses, and back radiation are most important. If waste-heat is to be discharged to waterways, these procedures will have to be used to predict the spatial and temporal distribution of the heat load and the resulting temperature rises.

It has thus been shown that we can expect increased amounts of heat to be dissipated in the future. We should now look at the effects of the increased heat discharges and then finally at the means available to ameliorate the consequences of these increased heat discharges.

Of all the effects of the increased heat to our streams, possibly the most striking is the induction of stratified flow by the discharge of these warmer waters to surface streams. It has already been indicated that, in 1965, water passing through the condensers of steam-electric plants rose 13° F. The most prevalent surface-water temperature in the coterminous U.S. is

55° F. Specific gravity at that temperature is 0.99942. With this 13° rise to 68° F., the specific gravity will fall to 0.99823. This density difference is sufficient to cause the water to flow as separate and distinct layers. In fact, underflows have been detected and maintained at half this density difference.

The return cooling waters could, of course, be discharged directly to the surface and would flow as a separate layer, or they could be discharged through diffusers at the bottom of the stream and mix completely with the flow in the cross-section and then flow as a unit through the entire cross-section. The recent report of the National Technical Advisory Committee to the Secretary of the Interior on water-quality criteria, however, rejects any discharges to the hypolimnion—that is, any heating of the hypolimnitic waters. This would also preclude artificial destratification. Reservoir surface waters may be 20° to 30° F. warmer than their bottom waters. In cases where the cooling water is derived from bottom waters of reservoirs further upstream, the condenser-cooling waters would have a density intermediate to both bottom and surface waters and would flow as an interflow.

As the water temperatures are raised, there is a lowering of the capacity of the water to hold oxygen. For example, the raising of the water temperature from 55° to 68° F. causes a loss of approximately 13 percent in the oxygen-carrying capacity of the water. Therefore, the amount of oxygen available under fully saturated conditions is less at elevated temperatures than at lower temperatures. As a result of the lower absolute amount of oxygen dissolved in the stream, for equal amounts of organic matter, the oxygen is depleted more quickly in a warmer stream.

If the heated discharge flows along the surface, then it is possible for reaeration to take place. However, if the heated discharge flows as an interflow, reaeration may not take place at all if this is below the thermocline. Consequently, organic material remaining in the bottom water after the oxygen content is exhausted will be stabilized anaerobically with noxious odors and the chemical changes associated with bottom sediments in a reducing environment. With a water-temperature rise, the metabolism of bacteria which break down the organic polluted matter is increased. Therefore, the oxygen demand is exerted at a greater rate. Consequently, the minimum oxygen concentration occurs closer to the discharge point. The deficit is aggravated, since there is less opportunity (time) for reaeration to take place. However, it is also true that reaeration rates are increased at higher temperatures. Metabolic rates are, however, increased at a greater rate.

Consequently, although the reaeration rate is increased, the net oxygen balance from considerations of reaeration and BOD exertion relative to the balance at 20° C. is smaller.

As the oxygen is depleted and reducing conditions occur, the ferric hydroxide which may have precipitated to the bottom is able to enter into solution and cause increased color in the bottom waters. Manganese chemistry in natural waters is similar to iron chemistry, and insoluble manganese forms under reducing conditions may enter into solution as manganous salts and cause even stronger colors than the iron solutions.

We can say that the addition of heat to a stream or reservoir is equivalent to the addition of an organic load, since it reduces the waste-assimilative capacity of a stream by reducing the amount of oxygen in solution and thereby available for oxidation of the organic material.

The rise in average water temperature tends to push the algal population in the direction of a greater percentage of blue-green algae. Some of the blue-green algae (*Anabaena, clathrocystis*), are notorious taste- and odor-producers in water supply systems. Consequently, more severe taste and odor problems are likely to result from a shift towards a higher percentage of blue-green algae. In addition, increased clogging of the filters may occur.

As the water temperature rises, and chemical reactions tend to proceed at a faster rate, the amount of chemicals required for water treatment is reduced. The estimated saving is 30 to 50 cents per million gallons of water treated per 10° F. rise. The cost of chemicals for treating water varies from $3 to $30, with an average cost of $14. The cost of chemicals at Chicago's Central District Filtration Plant is $6/million gallons. Therefore, by percentage, the savings are relatively small.

Water discharged from the hypolimnion of reservoirs may be too cold for efficient use as irrigation waters. Extreme and expensive designs have been used to permit warmer waters to be withdrawn. Particular attention has been paid to this problem for rice cultivation in California. Waste heat from power-producing stations could be used to raise river-water temperatures to the degree required for optimum growth conditions. Conversely, temperature rises of 9° to 19° F. have been recorded in the return flow from irrigated acreage. Such flows could therefore constitute an important thermal pollutant.

Waste heat from power plants could be used to keep shipping lanes through some of our interior waterways open for longer periods of the year, or possibly year-round. Calculations have been made to show that even a significant portion of the St. Lawrence Seaway could be kept open by a reasonable number of power stations installed along the St. Lawrence River.

This covers the chemical and physical aspects. Let us now turn to the most important, most controversial, and most emotional aspect, the biological effects of this increased heat load to our streams.

The major portion of the first National Symposium on Thermal Pollution concerned itself with the biological effects. From an engineer's point of view, it was quite disheartening because of the lack of solid information available upon which to base criteria, standards, and design. However, as one of the speakers, a biologist, suggested, the task of establishing criteria based upon the research done and funds available was somewhat like asking for a cure for cancer in one year with $100,000 available for research. Dr. Mount, Director of the National Water Quality Laboratory, put it more pointedly in his address when he said, and I quote, "Our present state of knowledge precludes the establishing of criteria for even the common water-quality parameters such as temperature, oxygen, and hydrogen ion concentration."

What do we know about the effects of changes in temperature upon aquatic life? We might start by quoting the comment in the book *Temperature and Aquatic Life,* "Since temperature is the most important single factor governing the occurrence and behavior of life, it not only affects the distribution of a single species, it may also modify the species composition of a community or of an ecosystem." Initial work on the effects of temperature concerned itself primarily with lethal factors, that is, the temperatures above which or below which specific species were not able to survive. In many cases, these two temperatures were quite close together. In recent years, more sophisticated analysis has shown that this information is insufficient to determine survival rates of species, because not only do individual species vary widely in their thermal tolerances, but also they vary according to their stage in life; they vary according to their previous acclimation to higher or lower temperatures; they vary in their resistance to thermal shock; and even though they may survive all of these changed conditions, the temperature rise may be sufficient to slow their responses so that they cannot capture adequate quantities of food material and, consequently, starve to death. Other possibilities are that the temperature change may be sufficient to prevent reproduction; or it may occur at the wrong time and trigger reproduction, only to have offspring die in the inhospitable environment.

Therefore, it is obvious that there are many possible ways where rises or falls in temperature may be sufficient eventually to wipe out the species, including necessary precursors in the food chain, even though members of the species may not all be killed immediately. It is also obvious that, in environmental situations rather than in laboratory tanks, the fishes or other aquatic organisms are subjected to a variety of stresses simultaneously. Laboratory tests where individual stresses are applied one at a time do not truly reflect what will occur in the stream, where some or all of the stresses are applied at one time. There are synergistic effects, as has been shown in many studies; for example, as temperature rises, the toxicity of certain substances also increases.

We might also ask why are effects of temperature on fish so acute? Primarily, we can say, because fish are poikilothermus, meaning that their body temperature follows changes in environmental temperatures rapidly, whereas in human beings there are a number of internal mechanisms which allow us to compensate for either higher or lower temperatures.

Some amplification of these generalizations seems desirable. Direct killing of fish may take place by four different mechanisms. They are:

An enzyme inactivity due to the saturation of the enzyme reaction

Coagulation of cell proteins

Melting of cell fats, and

Reduction in the permeability of cell membranes

We have mentioned that species vary in their thermal tolerances; but even within species, it has been shown for other stresses that individual members of species may vary as much as three times the average of the species. We might expect the same variation of the individuals from the mean in response to thermal effects. It has also been shown that thermal shock—that is, sudden and rapid changes in temperature—can be more harmful than continued exposure to a higher or lower temperature. Determinations of temperature tolerances in any case are difficult to evaluate, unless the holding temperature or the past thermal history of the fish is known. It has been shown that fish can acclimate themselves to higher or lower temperatures, but that the limit of this acclimation is not extensive. Springtime fish deaths have resulted from fish being subjected to warmer springtime waters after having been acclimated to colder winter waters. It has also been shown that it is usually easier to acclimate organisms to higher temperatures and at a greater rate than it is to acclimate them to lowered temperatures. It is also interesting to note that intermittent exposures to higher or lower temperatures for a sufficient number of hours per day can produce exactly the same acclimation effect as a continuous exposure.

Possibly the most important temperature information is that organisms require different temperatures during the various stages of their lives. The required temperature for reproduction may be quite different from the required temperature for hatching, for egg survival, for fry, and for activity and growth. It has been stated, with particular reference to marine and estuarine animals, that, with proper physiological conditions and enough room and food, the breeding season depends on temperature rather than salinity, light, pressure, or biological factors.

It is apparent, then, that future studies need to be a closer duplication of the environmental conditions which are obtained in nature rather than the laboratory studies conducted in aquaria. Mihursky refers to these as multi-varied studies. The problem, then, still remains: how to determine temperature requirements for all the varied organisms, including fish. Less then 5 percent of the 1900 fish listed in the American Fisheries Society as native to the United States and Canada have been examined for thermal effects. Reviewing these reports and those of thermal requirements of food-chain organisms, it strikes one as a gigantic crossword puzzle with a small corner or segment filled here or there, but with no over-all discernible pattern yet glimpsed. In modern language, we know some individual nodes, but have not yet learned the nature of the response surface. There are so many species, so many possible food chains, so many different stages in the growth of each of these various species that it would be well-nigh impossible to investigate each and every one of them.

Information that is presently available is insufficient to establish valid temperature criteria for each of these species at the present time. Consequently, Mount has suggested that we first determine the thermal requirements for the important species. By important species, he means those species and their supporting food chain which are desired and utilized by man. "The public," Mount states, "is not interested in protecting diatoms, may flies, or water fleas unless they know that such species are necessary to produce desirable shell- or fin-fish." Mount also makes the point that many people may get a great deal of pleasure from seeing a rare species of waterfowl or other birds, but not many people can get the same exhilaration out of seeing a rare species of algae. If this pragmatic approach were followed, it would, of course, simplify the work tremendously, and it would allow us to concentrate on the important aspects, to the public at least, first.

The recently issued report of the National Technical Advisory Committee to the Secretary of the Interior on Water-Quality Criteria emphasized the same approach, with recommended temperatures for various life stages of important commercial and sport fish. The committee states, "In view of the many variables, it seems obvious that no single temperature requirement can be applied to the United States as a whole, or even to one state; the requirements must be closely related to each body of water and its population." Therefore, maximum temperatures for various species of fish and for different stages of their lives, from egg spawning, egg development, and growth, are given. These provisional maximum temperatures, recommended as compatible with the well-being of various species of fish and their associated biota, range from 48° F. for spawning and egg development of some cold-water fish, such as Atlantic salmon, to 93° F. for growth of catfish and other warm-water forms. In addition, the recommendations for warm freshwater fish are that, during any month of the year, the temperatures should not exceed a 5° rise, and for cold-water fish, no heated effluent should be discharged in the vicinity of spawning areas. Inland trout streams, headwaters of salmon streams, trout and salmon lakes and reservoirs, and the hypolimnions of lakes and reservoirs containing salmonids should not be warmed. Neither should warm-water hypolimnions receive heated effluents unless this practice shall be shown to be desirable.

In essence, then, definitive information on thermal effects is not well known. However, sufficient data is known, for some important fish, to permit upper thermal limits to be set, with some justification.

We have now discussed the sources and magnitude of the thermal-pollution loading, and we have also discussed, to the extent of available knowledge, what the effects of this increased heat energy would be upon the biota of our streams and reservoirs. We now shall look at means of preventing this excess heat from going into the streams and reservoirs.

It is obvious from thermodynamic properties that the only other fluid which would have the specific heat, the available volume and price required to transfer the heat from the electricity-generating station to the environment is air. There are a number of means by which this can be done. We can divide these roughly into cooling ponds with manifold variations, such as spray-cooling ponds, etc., induced-draft cooling towers and natural cooling towers, and both wet and dry variations of both these classes, as shown in Table 2.

When heat loads were low, cooling ponds and spray-cooling ponds were very efficient, economically; but captive water bodies were not, necessarily, technically efficient means of venting this excess heat to the atmosphere. However, as heat loads became greater and land became more valuable, fewer land-intensive cooling devices were required, and forced-draft and natural-draft towers were developed. The driving force in the wet-cooling

TABLE 2. Types of Cooling Devices

Cooling Device	Area Required (Normalized)	Water Required Gal/kw/yr	Additional Unit Costs $/kw	Cost to year 2000 Billions
OPEN CYCLE		500,00		
POND	500–1000			
MECHANICAL DRAFT				
1. Wet	1–2	4000–8000	7	11
2. Dry		70	27	
NATURAL DRAFT				
1. Wet	1	4000–8000	11	16
2. Dry		70	25	60

SOURCE: "Cost to year 2000" after Richard N. Bergstrom, "Hydrothermal Effects of Power Stations." Paper read at ASCE Conference, May 1968, at Chattanooga, Tennessee, p. 9.

towers, is, of course, the evaporation of water, and the heat of vaporization is utilized to cool the rest of the water. Blow-down losses would be 2 percent to 3 percent of the circulating water. Over-all evaporative water losses are approximately 1 percent to 2 percent of the circulating stream, and in a water-short area, as some of the arid regions in this country are, this loss is cumulative and, over a highly developed stretch of the river, could be substantial. Some rivers in England have been reported to lose as much as 20 percent of their total flow through evaporation from cooling towers.

In arid regions, the use of dry-cooling towers appears to be quite attractive; and in England recently, dry-cooling towers based upon the so-called Heller System have been used with quite satisfactory results. The system was developed in Hungary and is also used in the Soviet Union. Dry-cooling towers reduce some of the problems that one has with wet-cooling towers, such as the concentration of chemicals, such as corrosion inhibitors, in the waters. When these chemicals are discharged to the streams, they themselves can prove to be a chemical pollutant in the water system. Primarily, however, the main advantage to the dry type is the approximate 8,000-fold decrease in water requirements in comparison to open-cooling cycles, and the 100-fold reduction in comparison to wet-cooling towers.

We might now look at some of the characteristics of cooling ponds, wet-cooling towers, and dry-cooling towers. The surface area required for cooling ponds is greater than that for induced-draft towers by a factor of about 500; and, surprisingly, induced-draft towers or forced-draft towers require more surface area than do natural-draft cooling towers in the larger sizes or if more than one is required at a site. Although each case has to be considered uniquely, as in all cases dealing with environmental problems, there are some rules of thumb that, in general, hold true among the three different types. The average cost of a cooling pond might even be less than once-through river cooling, if pumping costs are higher, etc. If we use once-through river cooling as the standard, according to Kolflat (U.S. Congress. Senate, 1968), the additional investment-capital cost required for induced-draft, wet-type tower is $7/kw; for induced-draft dry tower, $27/kw; for natural-draft wet type, $11/kw; and for natural-draft dry type, $25/kw. It is impossible, of course, to generalize completely on this matter, as each individual case is different.

We have looked at the costs of reducing thermal discharges. Though these cost figures are tenuous, the dollar value of the benefits achieved by this reduction are even more tenuous and difficult to estimate, as those of you who have tried to estimate the value of sport fisheries, recreation, etc., well know. Some ideas of the over-all cost of cooling-tower systems to the economy may be gathered from the estimates prepared recently by Bergstrom, who indicated that, by the year 2000, conventional wet induced-draft cooling towers with known technology will cost the nation $11 billion —$16 billion, if natural-draft is used; and $60 billion, if dry towers are used. It is, therefore, obvious that unless we find some auxiliary use for this low-grade heat that we are now discharging to the environment, or find new cycles that are thermodynamically more efficient, or accelerate technological developments that would make magnetohydrodynamics more immediately attractive, then we must be prepared over the next decades either to pay the price for the prevention of the heating up of our surface waters, to accept the consequences of heating these discharges and pay their toll, or else to work out some mixture of these limits.

It would appear, then, that our role as biologists and engineers is to determine the options available to us and the consequences of the options available to us to maintain water-quality standards and fisheries resources that the majority of the population desires. The final choice must be made by the body politic.

REFERENCES

Mihursky, J. A., and V. S. Kennelly. 1967. "Water-Temperature Criteria to Protect Aquatic Life." *American Fisheries Society, Spec. Publ.* (4):25.
Mount, Donald I. 1968. "Water-Quality Temperature Requirements for Aquatic Organisms." Paper read at National Symposium on Thermal Pollution: Biological Considerations, June 3–5, 1968, at Portland, Oregon. (See p. 00, *Biological Aspects*

of Thermal Pollution: Proceedings of the National Symposium on Thermal Pollutional, Portland, Oregon, June 3–5, 1968, edited by Peter A. Krenkel and Frank L. Parker, Nashville, Tenn.: Vanderbilt University Press, 1969.)

U.S. Congress. Senate. Committee on Public Works. 1968. *Hearings Before the Subcommittee on Air and Water Pollution: Thermal Pollution,* by Tor Kolflat. 90th Cong., 2nd sess., 6 February 1968, p. 54.

U.S. Department of the Interior. 1968. *Water-Quality Criteria: A Report to the Secretary of the Interior by the National Technical Advisory Committee, Federal Water Pollution Control Administration.* Washington, D.C.: U.S. Gov't. Printing Office.

Chapter **12**   Bruce A. Tichenor and William A. Cawley

RESEARCH NEEDS FOR THERMAL-POLLUTION CONTROL

PRECEDING papers of this symposium have dealt with all aspects of thermal pollution. The causes, effects, and control of heated waste-water discharges to surface waters have been discussed by national and international experts. A great deal of useful information has been disseminated; but unfortunately, large gaps exist in our knowledge concerning the control of this recently recognized water-pollution problem. Research is urgently needed to fill these gaps.

In discussing the research needed for thermal-pollution control, it is important to set priorities. Pollution-control research is action-oriented: that is, set up to solve present and future water-pollution problems, and priorities must be defined in terms of these problems; the more serious the problem, the higher the priority. Past speakers have spelled out the seriousness of present thermal-pollution problems and the potential for future pollution of this type. Research priorities for thermal-pollution control must now be set to control and prevent such pollution. Priorities can and do change, however, depending on such factors as technological advances and public opinion.

Priorities may be set on a national or local scale. For example, the thermal-power industry's use of over 80 percent of the nation's cooling water dictates that a sizable, high-priority research effort be directed at the discharges from these plants. However, in parts of the country, increasing stream temperatures and changing thermal regimes may not be due to heated-water discharges, but instead to other such man-induced factors as impoundments, streamflow depletions due to irrigation, and removal of shade trees by logging. In these areas, research on controlling thermal discharges would not be relevant to the problems, and thus would not receive high priority.

In this paper, we will spell out the research needed for better control of thermal pollution, and at the same time indicate the items of high national priority. It is emphasized that the priorities so indicated are based on current needs and are not intended to indicate the scientific merit of specific research items. The priorities have been set from a pragmatic, rather than a philosophical, point of view.

The research needs are segmented in three general categories:

1. Transport and behavior of heat in water
2. Treatment processes
3. Non-treatment controls

RESEARCH NEEDS

One important area of research need with respect to thermal pollution is that of biological effects. Dr. J. Frances Allen of the Federal Water Pollution Control Administration delivered a paper on this subject at the symposium's first session in Portland, Oregon, earlier this year. Her remarks will be published in the combined proceedings of the symposium's two sessions. Thus we shall not speak to this topic at this time.

Transport and Behavior of Heat in Water.

The nation's constant supply of water, coupled with the rising production of goods and thermo-electric power, means that an ever increasing heat load will be added to our waters. Therefore, it is essential that we understand the transport and behavior of heat energy in water. Much research has already been done to describe this transport and behavior, but much more work is needed. Mathematical and physical models of varying degrees of sophistication have been developed to determine the fate and persistence of heat in quiescent waters, flowing streams, estuaries, and the ocean. In broad terms, all of this work could be categorized as "Water-Temperature Prediction."

The ability to predict water temperatures accurately is necessary in order to determine the thermal impact of proposed waste heat discharges; changes in the hydraulic characteristics of a stream or other water body due, for example, to the construction of a dam; releases of water from stratified reservoirs with multi-level outlets, and unusual meteorological conditions. Water-temperature models can be classed as:

1. Macro: Involving a complete river or river system, lake or reservoir, estuary, or coastal area, and

2. Micro: Describing the distribution of heat in the immediate area of a thermal discharge.

To date, much of the work on both macro and micro models has been semi-empirical in nature, and the models are generally deterministic. High priority should be given to the development of temperature distribution models which are non-empirical and probabilistic.

The use of empirical coefficients—for example, in energy-budget computations—often leads to so-called generalized models whose coefficients are based on specific circumstances. The Lake Hefner (Oklahoma City) coefficient for evaporation, although adequate for computations on Lake Hefner, may have little relevance to reservoirs and rivers of different geographic and hydrologic characteristics. However, this empirical device is often used to predict evaporative heat losses in different parts of the country. The use of an evaporation equation based on mass transfer of water vapor within the turbulent boundary layer would not be dependent on empirical coefficients. This is just one example of the use of empirical factors in models. The use of empirical determinations for eddy diffusivity in thermal-plume computations is another example. The point is that research is required in these two cases, as well as other places, to eliminate empirical coefficients from generalized models. The theory is largely available to do this, and the advent of the large-scale digital computer makes possible the formerly untenable mathematical computations.

Most of the phenomena found in the natural environment are probabilistic: that is, we cannot predict exactly what a specific phenomenon will do; but we can make a statement about its distribution. For example, in predicting the temperature on a specific stream at location (x) and time (t), it is impossible to say with 100-percent confidence that it will be θ degrees Centigrade. However, using a probabilistic model with the proper distribution for the input variables, we could predict with some level of confidence, say 90 percent, that the temperature will be θ degrees Centigrade plus or minus some incremental temperature. Operations research techniques are available for dealing with the random or stochastic processes that exist in the real world. High priority should be given to developing water-temperature prediction models which use these techniques.

Another shortcoming of present water-temperature models is that most are based on simplifying assumptions which may or may not be true. Research is needed to treat rigorously all important aspects of the problem, thus eliminating such assumptions. Of all the assumptions made in such work, probably the most critical one is limiting the model to one, or, at most, two dimensions in space. High priority should be given to the development of rigorous three-dimensional, mathematical models for determining temperature distributions of complex hydraulic conditions. This work should be carried out on both the macro and micro scale.

The development of all heat-distribution models should be on a generalized basis. It now seems that each new situation demands a new model or modification of an old model, resulting in the development of a large number of models, each applicable to one specific situation. This is a tremendous waste of time, talent, and dollars. Instead of developing a new model for every slightly different problem, one generalized model should be readily available and used.

In research needs relating to the transport and behavior of heat in water, several specific problems come to mind. There is the problem of the prediction of the distribution of heat energy when heated water is injected into a stratified water body. More information is needed on what happens when a heated discharge is inserted in either the thermocline or hypolimnion of a stratified reservoir. How would such a discharge affect the reservoir's total temperature distribution? The question is important from at least two standpoints: A change in the vertical temperature distribution would change the volume of the epilimnion, the place where most of the biological activity in a lake takes place. Also, the temperature of the epilimnion would probably be changed. And a major change in the temperature distribution within the reservoir would affect the temperature of the water released at the dam. In terms of water resources management, this water temperature is important.

Therefore, methodology useful in predicting the effect of subsurface heated discharges into stratified water bodies is required. Again, this work should involve both the macro and micro scales.

Another problem is that design criteria for discharge structures should be based on more rigorous hydrodynamic considerations. At the present time, discharge pipes and diffusers are designed on the basis of model tests and empirical information. The design is often more art than science. Attention should be given to solving the theoretical equations of fluid motion for all types of discharge situations. These include single-point injection, as well as diffusers in the longitudinal, lateral, and vertical axes of the water body. Hydraulic situations differing in density gradients, velocities, turbulence, and boundary conditions should be accounted for.

The applicability of weather-station meteorological data for water-surface energy-budget calculations should be more rigorously determined.

The use of weather-station data for energy-budget calculations for large
aerial expanses is common practice today. Two questions need to be answered: Are the weather-station data representative of the conditions at or
near the water surface? Over what distances are the data from a weather
station applicable?

A method is needed to define the interfacial friction which occurs between a warm-water wedge and the underlying cold water. If heated water
is discharged on the surface of a lake, reservoir, or sluggish stream, it will
float on the surface. One factor to consider when determining the extent
of this warm-water wedge is the frictional force existing between it and the
colder water below. This frictional force is now estimated empirically and
research is needed to define its magnitude better.

Treatment Processes

Treatment of thermal wastes will be required to prevent further degradation and to promote improvement in present water quality. The public
demands that the most effective cooling methods be employed. Therefore,
high priority should be given to development of new and better cooling
devices for handling the ever increasing thermal loads.

The advent of large-scale thermal-power stations has resulted in huge
quantities of waste heat being produced at a single point. Although cooling devices, such as towers and ponds, have been used for many years,
only recently have such devices been required to dissipate the large quantities of waste energy now being encountered. The size, geometry, and hydraulic loading limits for both mechanical- and natural-draft towers are
not yet fully defined in terms of large thermal loadings. For example, will
it be necessary simply to increase the number of mechanical-draft units
proportionally to treat increasing quantities of waste heat, or would changes
in design result in fewer, more effective units? Also, is there a size limit
to natural-draft towers? Is it better to build one, or several? Will fan-
assisted, natural-draft towers lessen the need for additional units? Can dry
towers be made more economically? All of these questions need answers.

More work is also needed, on developing better design criteria for
cooling ponds. Here, again, the design is often based on questionable empirical evidence.

Problems associated with cooling devices should also be considered.
Blow-down from cooling towers, with its relatively high concentration of
dissolved solids, is a potential water polluter. The chemicals used to control
organic growth within the tower can concentrate in the blow-down and

cause problems. Another problem receiving much publicity but little investigation is the effect of cooling towers on local weather. Research is needed that can supply hard facts to the evaluators of potential tower locations with respect to the towers' impact on the local environment. At this time, very little scientific evidence is available on this matter.

Non-Treatment Control

If waste heat is not controlled by direct treatment, then other means must be used to prevent thermal degradation of our waters. Three non-treatment methods of thermal-pollution control should be considered: 1) use of waste heat, 2) water-quality management, and 3) more efficient methods of thermal-power production and/or new methods for generating electricity.

The most appealing method for preventing thermal pollution is the utilization of the waste heat, either before it enters the cooling water or before the heated cooling water discharges to the receiving water. Tremendous quantities of potentially useful energy are wasted because of the thermal inefficiencies of power production and other industrial processes. For example, a 1000-mw_e nuclear-power plant with an over-all thermal efficiency of 33 percent, a 5 percent in-plant loss, and an 80 percent plant-load factor, will discharge about 45×10^{12} (trillion) BTUs per year. In terms of industrial heat-energy requirements, this would have supplied about one-quarter of the heat energy needed by the textile industry, or one-twentieth of that required by the petroleum and coal products industry in 1962. Thus, the waste heat from the thermal generation of power could supply a major portion of the nation's industrial heat requirements. However, a single large thermal-power plant can produce more waste heat than can be effectively used by any single industrial plant. Only a very large industrial complex would satisfy the waste-heat load, but many technical and economic problems must be overcome before such use of waste heat would be possible on a large scale.

Waste heat is now being used in several instances to heat buildings and, according to *Chemical Week* magazine (May 25, 1968, p. 61), a Dow Chemical plant at Midland, Michigan, will use steam from a Consumers Power's 720-mw_e steam-turbine plant for process-heat requirements. However, on a national scale, such waste-heat utilization is very small, in proportion to the total load. Technical problems of power-plant design and heat transport must be solved. In case of steam use, it might be beneficial, from an over-all community standpoint, to reduce the efficiency of a power plant in order to supply economical heat to nearby users. This would represent a trade-off between electric power and steam use which could be optimized at the local level.

The use of the waste heat in the condenser discharge poses even a greater problem from an industrial point of view. Very few industrial processes can efficiently use energy for such low quality. However, agriculture is a prime contender for such energy use. Irrigating with heated water could promote faster growth and extend the growing season. Hothouses could be used to grow tropical or subtropical crops in the more temperate regions of the country. However, problems of soil adaptability, crop resistance to heat, and undesirable side effects, such as excessive growths of aquatic plants in irrigation canals, will have to be solved before the large-scale use of heated water for irrigation is a common practice.

High priority should be given to research for the development of methods for effectively utilizing waste heat for industrial and agricultural purposes.

Another potential use of condenser-discharge water is aquaculture. Marine and freshwater organisms may be cultured and grown in channels or ponds fed with heated water. For example, it may be possible to grow commercially valuable oysters in areas where they cannot normally survive due to low water temperatures. Power companies might be able to pay for required cooling devices with the profits received from aquaculture.

The need to manage our total environment has only recently been recognized, and we now realize that, to continue the effective use of our natural resources, environmental management is a necessity. In terms of thermal-pollution control, we can manage water temperature by effective use of regulated river systems. The release of cool, deep water from stratified reservoirs will permit the manipulation of the downstream thermal environment. Multi-level outlets will allow the selection of discharge water at different temperatures. Effective water-temperature prediction models, as discussed earlier, will permit the water resources manager to determine what the downstream temperatures will be, given various dam discharge temperatures. Successful completion of the research suggested on the transport and behavior of heat in water will provide the answers for implementing such a water-temperature control system.

The most direct method of decreasing the waste-heat load on the aquatic environment is to increase the thermal efficiency of electric-power generation processes. Modern coal-fired plants which operate at thermal efficiencies of about 40 percent are approaching the maximum efficiencies possible with the steam-turbine process. Nuclear plants of the boiling-water or pressurized-water types will probably not exceed thermal efficiencies of 33 percent, due to lower operating temperatures and pressures. Liquid-metal cooled reactors, gas-cooled reactors, and fast-breeder reactors will make possible higher efficiencies, but not greater than are now provided by coal-fired plants. Research is needed to develop new power-generation concepts which will result in higher thermal efficiencies. These new concepts may involve basic changes in the steam-turbine process as we know it today, or they may be completely new methods of power production, such as fuel cells or magnetohydrodynamics. Improvements in the gas turbine, which does not require cooling water, would also lessen the thermal burden on our nation's streams and lakes.

Non-thermal power-generation methods should also be considered. Horizontal turbines which use the kinetic energy of a stream's velocity instead of the potential energy of head, as in conventional hydropower installations, could extract vast amounts of energy from our nation's waters without producing waste heat. Pumped-storage plants can be used to produce peaking power. The generation of power using the ocean's tides, although now done on a small scale, has large potential.

A long-range research program is needed to develop the full potential of these thermal and non-thermal methods of producing power, but *now* is the time to start, so the power will be there when we need it.

SUMMARY AND CONCLUSIONS

In summary, the following four items have the highest priority for thermal-pollution control research:

1. Development of non-empirical, stochastic temperature-prediction models
2. Development of three-dimensional mathematical models for determining temperature distributions of complex hydraulic conditions on both macro and micro scales
3. Improvement of the effectiveness of cooling devices for handling large thermal loads
4. Development of methods for effectively utilizing waste heat, including industrial and agricultural use

Again, it is emphasized that these high-priority research items should not preclude work on other aspects of the thermal-pollution problem. For example, long-range, basic research needs should not be overlooked. The stated priorities should be used as a guide to indicate the areas of greatest research need.

In considering research needs, it is important to recognize that research involves several levels of activity:

1. Scientific advancement: basic research to develop new theories and concepts
2. Fundamental development: applied research, using desk-top studies and laboratory work to develop the technical feasibility of new ideas
3. Design and application: practical work on the pilot level to develop prototype applications of promising systems
4. Technology transfer: the dissemination of research findings to the appropriate people through written and oral communication

All these four stages of research are necessary; but the one most important, from the standpoint of those who are responsible for thermal-pollution control, is number four: *technology transfer.* Clear, concise reports on research findings enable the appropriate application to take place. Too often, a technically valuable piece of research work fails to reach the right people because of the way it is presented and/or disseminated. Care must be taken to transfer research findings effectively to those who must apply them.

Who is responsible for thermal-pollution-control research? Three segments of our society should each carry a fair share of the load: the regulatory agencies, who are responsible for maintaining and improving water quality; the polluter, who, by using water, impairs its use for other beneficial uses; and the water user, who benefits by having available more high-quality water. Concerted effort by all three will be necessary before the threat of thermal pollution is ended.

The wide variety of research needs mentioned in this paper shows the broad scope of the investigations necessary to solve the problem. Research to control thermal pollution will involve work in many fields not directly related to the typical water-pollution-control professions of chemistry, biology, and sanitary engineering. Economists will be required to evaluate the effect various cooling schemes may ultimately have on the public's pocketbook. Potential meteorological problems which may be associated with large cooling devices will require investigation by meteorologists and air-pollution specialists. Experts in various industrial processes requiring heat inputs will be involved in determining the feasibility of using waste heat from thermal-power stations for beneficial uses. Technical competence in agriculture will be called on to render professional opinions on the use of heated water for irrigation. Even sociologists have a part to play, in evaluating, for example, the impact of a large nuclear-power plant and associated cooling devices on the quality of life in the surrounding area. The point is that it will take an interdisciplinary research effort ultimately to control thermal pollution, calling for the participation of all relevant segments of the technical community.

DISCUSSION FROM THE FLOOR

Norman H. Brooks: I was interested in that last speaker's reference to the use of *empirical coefficients*. I believe, in many instances, he is referring to what I might call *physical coefficients*. Surely, he would grant the need for knowing such empirical information as the relationship of density to temperature of water, or the need for observations of turbulent mixing rates as they actually occur in nature. Much turbulence research, for example, in all fields, depends very strongly on experimental fluid mechanics. So I want to close, making the comment that experimental observations in hydraulics and fluid mechanics are just as important as Dr. Tichenor's particular approaches.

James W. Winchester: At this symposium, as well as the others I have attended, I believe one point comes through rather loud and clear: namely, thermal criteria cannot be the same for all rivers and all estuaries. There are some indications that the same criteria may be established for each estuary. I notice that we seem to have overlooked the idea of actually conducting research programs directed toward the efficient management of individual rivers and estuaries. All of us who are in the oceanographic sciences and the environmental sciences realize that each particular river and each particular estuary is somewhat of an individual problem. Maybe we should begin to think in terms of research programs directed toward understanding the dynamics of a particular estuary, or particular river, so that we can decide what should be permitted on a particular estuary, instead of trying to set general criteria. I mean, by management, such things as permitting amounts of heat in one particular estuary by permitting a different amount on some other estuary. I mean management in the real sense of regulating what waste goes into particular rivers or estuaries. Also, one might carry management further, to the point of time-phasing of releases in order to take advantage of the capability of

339

the environment to disperse and assimilate waste at some one time but not at another, instead of trying to establish the same criteria for all cases. Regulatory agencies now have the authority to enforce such management, but the dynamics of each system must be understood.

PARTICIPANTS IN INFORMAL DISCUSSIONS

PETER ACKERS
  Senior Principal Scientific Officer
  Hydraulics Research Station, Wallingford, Berkshire, England

ALLEN F. AGNEW
  Director, Water Resources Research
  Center, Indiana University, Bloomington, Indiana

NORMAN H. BROOKS
  Professor of Civil Engineering,
  Keck Laboratories, California Institute of Technology, Pasadena,
  California

ROBERT S. BURD
  Director, Water-Quality Standards,
  Office of Program Planning and Development, Federal Water Pollution
  Control Administration, Washington, D.C.

R. S. CARTER, JR.
  Atomic Energy Commission, Aiken,
  South Carolina

WILLIAM A. CAWLEY
  Acting Director, Pollution Control,
  Technology Branch, Division of Research, Federal Water Pollution
  Control Administration, Washington, D.C.

MILO A. CHURCHILL
  Chief, Water-Quality Branch, Tennessee Valley Authority, Chattanooga, Tennessee

WILLIAM SPENCER DAVIS
  Springfield, Virginia

H. R. DREW
  Texas Electric Service Company,
  Fort Worth, Texas

EDWARD E. DRIVER
  Research Engineer, Tennessee Valley Authority Engineering Laboratory, Norris, Tennessee

JOHN E. EDINGER
  Assistant Professor, Sanitary and
  Water Resources Engineering, Vanderbilt University, Nashville, Tennessee

REX A. ELDER
  Director, Engineering Laboratory,
  Tennessee Valley Authority, Norris,
  Tennessee

CARLOS FETTEROLF
  Chief, Water-Quality Appraisal Unit,
  Michigan Water Resources Commission, Lansing, Michigan

JOHN FOERSTER
  Marine Research Laboratory, University of Connecticut, Noank, Connecticut

GORDON L. FORD
  Senior Mechanical Design Engineer,
  Tennessee Valley Authority, Knoxville, Tennessee

RICHARD F. FOSTER
  Pacific Northwest Laboratories, Battelle Memorial Institute, Richland,
  Washington

KEITH FRY
  International Paper Company, New
  York, New York

THOMAS P. GALLAGHER
  Southeast Water Laboratory, Athens, Georgia

341

FRANCIS E. GARTRELL
Assistant Director of Health, Tennessee Valley Authority, Chattanooga, Tennessee

H. J. GORMLY
Pacific Gas and Electric Company, Berkeley, California

EMORY H. HALL
Bonneville Power Administration, Portland, Oregon

G. EARL HARBECK, JR.
Research Hydrologist, U.S. Geological Survey, Denver, Colorado

DONALD R. F. HARLEMAN
Professor of Hydraulics Engineering, Hydrodynamics Laboratory 48-213, Massachusetts Institute of Technology, Cambridge, Massachusetts

H. A. HAWKES
Department of Biological Sciences, University of Aston at Birmingham, Gosta Green, Birmingham, United Kingdom

S. LEARY JONES

WILLIAM L. KLEIN
Ohio River Valley Water Sanitation Commission, Cincinnati, Ohio

JOHN W. LEBOURVEAU
New England Electric System, Boston, Massachusetts

ROY E. NAKATANI
Director, Ecology Section, Pacific Northwest Laboratories, Battelle-Northwest, Richland, Washington

LORING F. OEMING
Executive Secretary
Michigan Water Resources Commission, Lansing, Michigan

SAM POSNER
Food and Drug Research Laboratory, Inc., Maspeth, New York

MACK S. PRICHARD
Tennessee State Department of Conservation, Nashville, Tennessee

GEORGE H. RAND
International Paper Company, New York, New York

W. R. SHADE
Chief Mechanical Engineer, Gilbert and Associates, Reading, Pennsylvania

EDWARD SILBERMAN
Director, St. Anthony Falls Hydraulic Laboratory, Minneapolis, Minnesota

ALEX F. SMITH III
Assistant to Chief Engineer, Gilbert and Associates, Reading, Pennsylvania

VERN W. TENNEY
San Francisco, California

LARRY B. VAUGHAN
KDI Corporation, Cincinnati, Ohio

JOHN C. WARD
Associate Professor of Civil Engineering, Department of Civil Engineering, Colorado State University, Fort Collins, Colorado

CHARLES WASELKOW
Manager of Production, Public Service Company of Colorado, Denver, Colorado

LEON W. WEINBERGER
Vice-president, Zurn Industries, Washington, D.C.

EUGENE B. WELCH
Supervisor, Biological Section, Health and Safety Division, Tennessee Valley Authority, Chattanooga, Tennessee

JAMES W. WINCHESTER
Santa Barbara, California

TOM WRIGHT
New York State Department of Agriculture, Cornell University, Department of Conservation, Ithaca, New York

WALTER O. WUNDERLICH
Tennessee Valley Authority Engineering Laboratory, Norris, Tennessee

INDEX

Acclimation temperature, 42, 65
Advection, 125
Aeration: in cooling towers, water chemistry, 51
Air bubbles: used to deflect fish, 228
Air pollution, 275, 302
Air-water interfaces, 165
Alevins. *See* Fry
Alewife: effect of water temperature on spawning, 43
Algae: found in water at 73° C., 20; thermal optima, 26; composition and succession of algal populations, 47; water temperature optima, 47
Algae, blue-green. *See* Cyanophyta
Ambient turbulence, 180
American Fisheries Society, 324
American mollusks. *See* Mollusks, American
American shad. *See* Shad, American
Ammonia: toxicity to fish, 41; nitrification, 52; increase with water temperature increase, 60
Anaerobic microbial activity, 52
Approach: in cooling-tower operation, 273
Aquaculture, 335
Aquatic life: effect of changes in temperature, 4, 47, 322; oxygen consumption, 4; thermal tolerance, 22
Atmosphere: as a heat sink, 303
Atmospheric cooling, 124
Atmospheric heat dissipation: surface, 156
Atomic Energy Commission: and thermal pollution, 11, 12, 13
Aust Nuclear Power Plant (United Kingdom), 184
Autecology, 18
Automation: at Wallingford Hydraulics Research Station, 205–210

Bacteria: autotrophic, 17; found in water at 85° to 88° C., 19; optimum thermal range, 20; temperature tolerances in hot springs, 20; effects of heated discharges, 49–50. *See also* Micro-organisms
Baffle: used to straighten canal flow, 221
Baffle wall. *See* Skimmer wall
Barkley Dam, 215
Barkley Lake, 215
Bass: largemouth, 31; potential effects of heated water on, 65, 75; maximum water temperature, 75
Bay-lake cooling system, 305
Beautor Power Plant (France), 113, 119, 120, 121
Benefit-cost ratio, 78
Benthos, significant, in rivers, 47–48
Berkeley Nuclear Power Plant (United Kingdom), 184, 185, 186
Biochemical Oxygen Demand (BOD): sewage effluent in cooling towers, 255, 262
Biological indicators, 30
Biological oxidation: increase as water temperature increases, 51
Biscayne Bay, Florida, 245–247 *passim*
Black Rock Nuclear Power Plant (United Kingdom), 184, 189–191
Black Rock outfall/intake model, 189–191, 220
Blow-down: from cooling towers, 308, 333, losses, 326
Bluegills. *See* Sunfish
Blue-green algae. *See* Cyanophyta
BOD. *See* Biochemical Oxygen Demand
Body fats: in determining upper lethal temperature, 22
Bowen ratio, 126, 135, 139
Brook trout, 36, 42

Browns Ferry Nuclear Power Plant (Alabama), 64, 67, 79, 82, 94, 170, 173, 244, 245, 316
Buoyant plume, 180, 181

Canals: cooling in, 119: Leaburg Canal, 225
Carbon dioxide, 52
Carnivores, 17
Carnot cycle efficiency, 317
Carp, 4, 40
Catfish, 65, 75, 325
Cell membrane: changes in lipid structure, 22
Certification: relative to heat from power plants, 12–13
Cheatham Dam, 215
Cherokee Reservoir, 137
Chezy coefficient, 213
Chironomids. See Diptera
Chlorides, 52
Chlorine, 43
Chromates, 308
Chub, 41
Circulating pumps, 112
Coastal Power Plants, 245
Colbert Steam Plant (Alabama), 81
Cold water, 273
Colorado City, Lake (Texas), 135, 308
Condenser-cleaning practices, 310
Condenser water. See Cooling water
Conduction, 126
Connecticut River, 6, 69, 174
Construction: biological-limnological consultants recommended, 90
Consumers: in an ecosystem, 17
Convective spread, 179, 180, 187
Continuous readings, 122
Cooling: of rivers, degree of, 113–114
Cooling, air: direct, thermodynamic problems, 303–304
Cooling ponds: widely used, 7–8; theory of, 134–136; system, 306; better design criteria needed, 333; mentioned, 325, 326
Cooling processes, 125
Cooling systems: comparative costs, 304–307
Cooling towers: corrective device for pollution, 7–8; aesthetic pollution, 8, 107; world's largest, 8; mechanical-draft, 8, 249–281 passim; natural-draft, 8, 249–281 passim; in Great Britain, 52; difficulties in planning, 108; costs, 246, 279, 280, 288, 321, 327; water requirement in Colorado, 251; cross-flow versus counter-flow,

252, 253, 273; dry versus wet, 253; proximity to transmission lines, 255; mist eliminators, 256; icing and de-icing, 257, 258; operation, 257–259; wood deterioration, 259–260; public relations, 260–263; fog (vapor), 262, 263, 276; glossary for, 273; effect of wind velocity, 274; mechanical-draft, specification figures, 274; evaporation, 276; windage losses, 284; natural-draft, run-of-river makeup, 305; natural-draft, reservoir makeup, 306; blow-down from, 308, 326, 333; effect on local weather, 334; mentioned, 325–326
Cooling water: factors limiting use, 5; decrease in requirements by technological controls, 7; intake, 165; discharging, 165; percentage of higher temperatures added by nuclear power plants, 244; need for, projected to 1978, 244; economic sources becoming exhausted, 254; power-plant requirements, 254; use of sewage effluent, 254, 255; discharge temperature criterion, Green River, 276; costs, 282–301; industrial use of, 282, 313
—condensers: cooling-water discharges, 125; stratification in condenser cooling water, 144–163; complete mixing, 157
Copepods, 26
Cowlitz River, 226
Criteria: determination of, for water-temperature standards, 63–68
Critical or minimum depth, river flow, 154
Crustaceans: multiplication rate increased, 26
—freshwater: mutation of, 23; temperature tolerance, 45
Crystal River (Florida) coastal power plant, 245
Cumberland Steam Plant (Tennessee), 215
Currents, water: importance in thermal discharges, 111
Cyanide, 41
Cyanophyta: temperature tolerance in hot springs, 20; seasonal successions temperature-induced, 27; water-temperature optima, 47; as cause of taste and odor problems, 321

Dace: oxygen requirements, 40
Dams: Wheeler Dam, 64; Barkley Dam, 215; Cheatham Dam, 215

Darcy-Weisbach equation, 218
Decomposers, in an ecosystem, 17
De-icing, cooling towers, 257, 258
Delaware River, 44, 46, 48
Densimetric Froude number, 147, 156, 157, 159, 174, 185
Density stratification, 144–146
Desalination: use for waste heat, 9
Desmid (*Staurastrum*), 27
Diatometer, 48
Diatoms: *Navicula ambigua*, 15; *Asterionella formosa*, 27; *Fragilaria crotonensis*, 27; water-temperature optima, 47
Diffuser, multi-port, 94; design, 170
Dilution, wastes. *See* Waste dilution
Dinoflagellates: *Ceratium hirundinella*, 27
Diptera: chironomid larvae, 46; effect of shock temperature increases, 47
Discharge canal: with skimmer wall to prevent recirculation, 215, 216
Discharge of cooling water, 165
Discontinuous measurements, 122
Dispersion, 125
Diversity indexes, 69
Drift eliminators, 275
Drift loss, cooling towers, 255–257, 275
Dry-bulb temperature, 273
Dublin Harbor model studies, 196, 198
Dutch East Indies: temperature tolerance in thermal springs, 20

E. Coli: effect of heated discharges on, 49
Ecological changes: induced by discharge of heated waters, 15–57
Ecology of fresh waters, 16
Ecosystem: concept of, 17–19; aquatic, role of micro-organisms, 28; aquatic, temperature factor, 28
Ecotypes, 26
Eggs: LT50 values, 40; effect of water temperature on, 65, 66
Electricity, 289–292
Electric power: and thermal pollution, 5; percentage produced by fossil fuel, 6
Electric power plants: cooling-water requirements, 254, 286; greatest source of heat added to waterways, 313; rate of increase, 314; Konin Electric Power Plant, 49
Electrogasdynamics: and pollution, 7
Energy budget: to determine fate of excess heat, 319; calculations, research needed, 332; mentioned, 135
Enforcement of standards, 101

Enteric bacteria: effect of heated discharges on, 49
Entrainment, 124, 157, 178
Environment: destruction by uncontrolled pollution, 248
Enzymes: thermal inactivation, 22
Ephemeroptera. *See* Mayflies
Epilimnion: effect of heated water on, 332.
Equilibrium temperature, 119, 126, 127, 131, 134
Estuaries: ecology altered by temperature change, 4; direction of current reverses with tides, power-plant cooling, 111; modeling, 181; flow, recirculation paths, 182; stratification in, 184; recirculation paths of flow, 215
—named: Severn, 184, 189, 210; Forth, 199, 202
Eurythermal, 19
Evaporation: from cooling towers, 276; economic consequences, 283, 286, 295–296, 303; importance, 303; from reservoirs when heat is added, 308–309; mentioned, 126, 179, 181
Excess heat. *See* Thermal discharges
Experimental verification, 153, 177–224 *passim*

Federal Power Commission National Power Survey, 6
Federal Power Commission: average heat rates, 7; and fossil-fuel plants, 11
Federal Water Pollution Control Act, 10, 87
Federal Water Pollution Control Administration, 314
Field data. *See* On-site data collections
Fish: dependence on temperature changes for migration, spawning, 4; effect of increased temperatures on, 18, 31, 40; LT50 values of adults, 40; temperature tolerance, 40; oxygen requirements, 40; responses to temperature differences, 41–45; migratory, effect of temperature, 42–43; effects of heated water discharges, 43–45, 67; deflecting away from power plants, 225–242; four mechanisms of killing, 323. *See also* names of species
Fish screens. *See* Screens, water
Flatworms: *Dugesia lugubris*, 15; thermal limitations, 25; effect of shock temperature increases on, 46
Flow: critical or minimum depth, rivers, 154
Flow-recorder, 207

Flow stability, 148–150

Flumes: Troy test flume, 235

Fog: from cooling towers, 262, 263, 276

Food chains: affected by warming coastal waters, 176; mentioned, 69

Fort Martin Station, Allegheny Power System, 265

Fort St. Vrain Nuclear Power Plant (Colorado), 271

Forth Estuary, 199, 202

Fossil-fueled power plants: heat waste, 5; percentage of thermally-generated electricity produced, 6; greatest users of water for cooling, 244; cooling-water requirements, 254, 286

—named: Paradise Steam Plant (Kentucky), 59, 272, 274, 276, 280, 312; Widows Creek Steam Plant (Alabama), 80; Colbert Steam Plant (Alabama), 81; James H. Campbell Electric Power Plant (Michigan), 88; Montereau Power Station (France), 112; Vaires-sur-Marne Power Station (France), 112–113, 118, 120, 121; Beautor Power Plant (France), 113, 119, 120, 121; John Sevier Steam Plant (Tennessee), 137; Shawville Power Plant (Pennsylvania), 139; Pigeon House Power Station (Ireland), 196, 198; Ringsend Power Station (Ireland), 196, 198; Longannet Power Station (Scotland), 199–202; Kincardine Power Station (Scotland), 202; Heysham Power Station (England), 203, 205; Cumberland Steam Plant (Tennessee), 215; Crystal River Coastal Power Plant (Florida), 245; Turkey Point Coastal Power Plant (Florida), 245, 246, 247

France: riverside thermal power plants, cooling, 110–123

Freshwater ecology, 16–19

Freshwater fisheries, 75

Freshwater runoff: projected need as cooling water by 1978, 244; by 1980, 282

Frictional force, 333

Froude Law, 179, 180

Froude number: in model design, 177, 185; definition, 187; Froude model, 213; densimetric, 147–159 *passim;* mentioned, 146–152 *passim*

Fry: upper-thermal limits of newborn, 22

Fungi, 49–50

Fusion. *See* Thermonuclear fusion

GEA Air-Cooled system, 264, 266, 267, 269

Glut herring, 43

Goldfish, 40

Grande Ronde River, 235

Gravity, 177

Great Britain. *See* United Kingdom

Green River, 59, 61

Gross National Product, 314

Heat balance: oceans and reservoirs, 166

Heat budget, 127

Heat budget equation, 214–215

Heat discharge. *See* Thermal discharges

Heated water. *See* Thermal discharges

Heat exchange, rate of, 126

Heat flux, 158

Heat sink: atmosphere, 303; oceans, 311

Heat transfer: water to atmosphere, 224; temperature differential required, 303

Hefner, Lake, 135

Herbivores: in an ecosystem, 17

Heysham Power Station (England), 203, 205

High temperature gas-cooled reactor: Fort St. Vrain Nuclear Station (Colorado), 271

Holston River, 137–138, 139

Hong Kong power stations and desalination plant, 191–196

Hot springs: temperature tolerances for micro-organisms, 20

Hot water, 273

HTGR. *See* High temperature gas-cooled reactor

Hydraulics, 170, 181

Hydraulics Research Station (United Kingdom). *See* Wallingford Hydraulics Research Station

Hydrogen ion concentration: in cooling towers, 52

Hypolimnion: of reservoirs, 75, 321, 332; NTAC report, 320

Icing: in cooling towers, 257, 259

Indicators: benthic macro-invertebrates, 47

Inertia, 177

Infrared radiation: airborne, 210

Ingersoll's heat-flow theory, 138

Injunction, 101

Inlet wet-bulb temperature, 253

Insects: nymph immobilized by cold air, 4; water-temperature tolerance, 46. *See also* names of species

Instrumentation: at Wallingford Hydraulics Research Station, 205–210; mentioned, 114
Intake, cooling water, 165
Interconnections, 312
Interfaces, 151, 156, 218
Interfacial friction, 333
Interior, Department of: and thermal plant pre-construction review, 11, 12; and temperature standards, 75–76
Interior, Secretary of the. See Udall, Stewart
Internal hydraulic jump, 155
Irrigation: as use for heated water, Pacific Northwest, 8

James H. Campbell Electric Power Plant (Michigan), 88
Jet diffusion, 168
Jet mixing, 165
Jet ports, 160, 162
Jet zone: in Hong Kong model studies, 194, 195; modeling, 194, 195
Jets, heated water, 157
Jets, buoyant: discharged horizontally, 168; fluid mechanics of, 168; in dispersal of heated water, 178–179; mentioned, 166
Jet diffusion, 179–181
John Sevier Steam Plant (Tennessee), 137
Jump: internal hydraulic, 155

K coefficient, 120, 121, 126
Kincardine Power Station (Scotland), 202
kLx, 120
Konin Electric Power Plant, 49

Lake cooling, 252, 259
Lake trout, 36, 75
Lakes, Polish, 69, 70
Lakes: standards and mixing zones, 106; water-temperature data, 128; ecological impact of small surface-warm area, 165; water requirement by lake-cooling in Colorado, 251
—named: Windermere, 25; Lichen, 48, 49; Slesin, 49; Colorado City, 135, 308; Hefner, 135; Barkley, 215; Valmont, 309
Largemouth bass. See Bass
Larval growth stage: and water temperature, 65, 66
LC50. See Median lethal concentration (LC50)
Leaburg Canal, 225

Legislation: for pre-construction-review authority, thermal plants, 11
Lethal temperature, 62
Lichen, Lake, 48, 49
Liffey River (Ireland), 196, 198
Lock-exchange problem, 187
Longannet Power Station (Scotland), 199–202
Los Angeles: waste heat and solar radiation, 98; ocean outfall schematic, 167
Louver system: for deflecting fish, 226; installed at Tracy, California, 235
Lower median lethal temperature, 36
LT50. See Median lethal temperature (LT50)
Lx, 120

McKenzie River, 225
Macro-invertebrates, temperature tolerance, 45–46
Magnetohydrodynamic power plants: at Arnold Engineering Development Center (Tennessee), 318; in Moscow, 318
Magnetohydrodynamics: and pollution, 7
Marine waters: temperature rise, 75
Marne, River, 113, 118
Mass transfer, 135, 136
Mass transport, 180
Maximum lethal temperature, 32
Maximum tolerable temperature, 32–35
Mayflies, 25
Median lethal concentration (LC50), 35
Median lethal temperature (LT50), 35
Median tolerance limit (TLm), 32, 35
Messinger ratio, 139
Meters. See Instrumentation
Metric units: use in calculations, 171
MHD systems. See Megnetohydrodynamic power plants
Michigan, state of: defines unlawful pollution, 86–87; water-quality standards of, 86–87, 90
Micro-organisms: optimum thermal range of, 20; temperature tolerances in hot springs, 20; effects of raised temperature on, 47. See also Bacteria
Midges: thermal tolerance, 45
Miniature Propeller Current Meter, 207–208
Minnows, 41
Mist eliminators: in cooling towers, 256
Mixing, 170, 173
Mixing zones: water-quality standard, 10, 73, 76, 79–81, 86, 93, 94, 100, 73–109 passim; in high-quality waters, 100

Model studies: two-dimensional undistorted, diffuser pipe, 162; schematic of ocean outfall, 167; heated-water discharges, 177–212; modeling principles, 179; comparison with field data, 182; distorted scale, 182, 194, 202, 217; natural scale, 182, 217; Black Rock outfall/intake model, 189–191, 220; Hong Kong power stations and desalination plant, 191–196; recirculated water, 194–198; for less than full tidal cycle, 196; Dublin Harbor, 196, 198; of variation of salinity on a vertical, 198; outfall design, 198; Longannet (Scotland), 199–202; Heysham (England), 203, 205; research needed, 330–332, 336
Mollusks, American: temperature tolerance, 45
Momentum input, 168
Monitoring: water temperature standards, 76
Montereau Power Station (France), 112
Multiple-jet diffusers, 166
Muskie, Senator Edmund S., 11, 280
National Power Survey, 6
National Technical Advisory Committee: water-quality standards recommendations, 74, 75, 76, 102, 103, 104, 106; report on water-quality criteria, 320
Nitrate concentration, 52
$NO_3:PO_4$, 27
Non-degradation policy, 100
Non-treatment control, 334–336
NTAC. See National Technical Advisory Committee
Nuclear-industrial parks, 9
Nuclear power plants: heat wastes of, 5; number being built and planned, 6; riverside, cooling, 110–123; percentage of thermally-generated power produced by, 244; add higher temperatures to cooling water, 244; cooling-water requirements, 254; rise in costs, 316
—named: Browns Ferry (Alabama), 64, 67, 79, 82, 94, 170, 173, 244, 245, 316; Aust (United Kingdom), 184; Oldbury (United Kingdom), 184; Tidenham (United Kingdom), 184; Berkeley (United Kingdom), 184, 185, 186; Black Rock (United Kingdom), 184, 189–191; Fort St. Vrain (Colorado), 271; Oyster Creek (New Jersey), 311
Nuclear reactors, 314

Nutrients: and algal growth, 27; in seasonal phytoplankton successions, 28; increased recycling, 63
Nysa Luzycka River, 48

Ocean outfall, schematic, 167
Oceans: special diffusion structures, 166; thermal stratification, 166; avoiding thermal pollution in, 171; as available heat sinks, 311
Odor: and high concentration of blue-green algae, 321
Ohio River Valley Water Sanitation Committee: water-quality criteria of, 74
Oise, River, 115, 121
Oldbury Nuclear Power Plant (United Kingdom), 184
Oligothermal, 23
On-site data collections, 182
Organic pollution, 28
ORSANCO, 108
Osmo-regulation, 22
Oxygen: consumption by aquatic vertebrates, 4; and upper thermal limits, 22; requirements of fish, 40; concentration in water lowered as water temperature rises, 320, 321
Oyster Creek Nuclear Power Plant (New Jersey), 311
Oysters: warm-water cultivation, 8

Pacific Coast: surface discharge off, 124
Paradise Steam Plant (Kentucky), 272, 274, 276–280
Partial-arc supports, 252
Penalties: for violation of water-quality standards, 101
Perch: oxygen requirements, 40. See also Yellow perch
Periphyton: significant, in rivers, 47–48; response to raised water temperature, 59; production of organic matter, 61
pH. See Hydrogen ion concentration
Phenols: threshold of toxic concentration, 41
Phosphates, 60
Photoperiod, 65, 66
Photoperiodism, 40
Phytoplankton: seasonal cycle, 26; effect of heated water on, 48–49
Pigeon House Power Station (Ireland), 196, 198
Plankton: seasonal successions, 26, 28. See also Phytoplankton
Platte River, 260, 261

Poikilothermal: organisms, 19; adaptation, 22–23; fish, 323

Poisons, 41. *See also* Toxicity

Pollution: definition of, 3; aesthetic, 8, 107

Polythermal, 23

Ponds: aerated, to control waste heat, 246

Pralgometer, 48

Producers: in an ecosystem, 17

Productivity: of heated lake, 49; primary, 69, 70

Prototype tests: verification, 213–214, 222

Protozoa: temperature tolerance in hot springs, 20

Public Service Company of Colorado, 249, 250, 251, 255–263 *passim,* 271, 274

Pumpkinseed (fish), 31

Pumps: circulating, in French power plants, 112

Radar screen, 209

Radiation: and Revelle's statement, 98; short- and long-wave, 214; mentioned, 126

Radioactive tracers. *See* Tracers

Rainbow trout: oxygen requirements, 40; toxicity of cyanide, 41

Range: in cooling-tower parlance, 273

Ratemeter, instantaneous, 208

Reaeration: in cooling towers, water chemistry, 52; occurrence with heated discharges, 320, 321

Recirculated water: control by skimmer wall, 145, 153; at intake, 158; in model studies, 194, 198; prevention with skimmer wall and discharge canal, 215; recirculation cooling systems, 284; costs, 289–294; mentioned, 111, 120, 121, 182

Reclamation, Bureau of, Tracy, California, 235, 238

Redwood: fill, in cooling towers, 259

Relative humidity, 273

Reproduction: effect of water temperature on, 65, 67

Reservoirs: special diffusion structures, 166; ability to dissipate added heat by evaporation, 308–309

—named: Wheeler Reservoir, 64; Cherokee Reservoir, 137

Respiration: temperature-dependent, 20

Revelle: on radiation and energy, 98

Reynolds number, 179–181 *passim*

Ringsend Power Station (Ireland), 196, 198

River pollution, definition, 29

Rivers and streams: ecology altered by temperature change, 4; effect of temperature on oxygen consumption of aquatic vertebrates, 4; thermal discharges and navigation, 15, 283; effects of heated discharges on microbiology, 50; temperature cycle, 114; importance of rate of flow, 117; daily water-temperature cycle, 128; critical or minimum depth for lower-layer flow, 154

—named: Connecticut, 6, 69, 174; Delaware, 44, 46, 48; Thames, 46; Schuylkill, 47; Nysa Luzycka, 48; Vistula, 48; Seine, 48, 49; Green, 59, 61; Tennessee, 63; Marne, 113, 118; Oise, 115, 121; Susquehanna, 134, 138, 139; Holston, 137–138, 139; Liffey (Ireland), 196, 198; McKenzie, 225; Cowlitz, 226; Grande Ronde, 235; Platte, 260, 261

Riverside thermal power plants, France, 110–123

Roach: lethal temperature for, 22; oxygen requirements, 40

Run-of-river cooling system, 111, 305

S. 3206, Section 5, 11

Saline wedge: arrested, 149

Salinity, 184, 198

Salmon: spawning prevented by too-warm water, 4; maximum water temperature for, 75; behavior, 95; steelhead and chinook, 222; Atlantic, maximum water temperature for spawning, 325

Salmonids: upper-thermal limits of eggs, 22

Schuylkill River, 47

Screens, water, 225–242

Sea lamprey, 45

Seas. *See* Oceans

Sea water: for cooling, 171

Seine, River, 48, 49

Severn Estuary, 184, 189, 210

Sewage: effluent for cooling-tower makeup, 254, 255

Sewage fungus: effect of heated discharges, 50

Sewage outfall, 166

Shad, American, 45

Shawville Power Plant (Pennsylvania), 139

Shock temperature, 46

Silica, 27

Simulation. *See* Model studies

Skelford's law of tolerance, 19

Skimmer wall: control of condenser-water recirculation, 145; size of opening, 153, 156; theoretical relationships, 153; control of recirculated water, 153, 159

Slesin, Lake, 49

Smolt: salmon, toxicity of cyanide, 41

Snails: *Physa acuta,* 15

Solar radiation: ultimate disposal of waste heat, 97; transports heat across water surface, 126

Sound: as fish-guiding device, 228

Southwestern Public Service Company (Texas), 249, 250, 251, 253, 254, 274

Stenothermal, 19, 23

Stephan-Boltzmann constant, 126

Stoneflies, 45

Stratification: of condenser-water discharge, 144–163; of the ocean, thermal, 166, 168; Severn Estuary, 184

Stratified flow: theory, 146; profile, 152; thermal stratification, 157; caused by increased heat in streams, 319

Stream-bed communities, 30

Streams. *See* Rivers and streams

Streptococci: effect of heated discharges on, 49

Striped bass, 45

Submarine outfall, 166

Submergence: Los Angeles and Los Angeles County, 175; Puget Sound, 175; San Diego, 175

Sulphates: concentration, 52

Sunfish: effects of heated water on, 31, 65

Susquehanna River, 134, 138, 139

Surface-heat exchange. *See* Heat exchange

Surface-heat transfer coefficients, 174

Surface spreading, 165, 170

Surface spreading: versus completely mixed, 93, 95

Synecology, 18, 26

Taste: problem increases with high percentage of blue-green algae, 321

Temperature: as ecological factor, 19–31

Temperature and Aquatic Life, 322

Temperature criteria, Great Lakes states, 85

Temperature-control system, 209

Tench: oxygen requirements, 40

Tennessee River, 63

Tennessee Valley Authority: lakes, 214; engineering laboratory, 215

Tennessee Valley Authority Project: experiments on heated-water discharges, 66–68; mentioned, 58, 59

Thames River, 46

Thermal death, 20–22

Thermal death point, 32, 36

Thermal discharges: technological controls, 7–8; need for productive use, 8; possible navigation aid, 15, 321; surveys below, 46–47; TVA experiments, 66–68; disposal problem, 124; and warm-water wedge thickness, 163; and multiple-jet diffusers, 166; dispersal, 178; controlling, 246; economic considerations, 247, 282–301; need for direction and control, 281; quantities, 283; power plants as greatest contributors of, 313; control, research needs, 329–330; non-treatment control, 334–336. *See also* Thermal pollution

Thermal expansion, 184

Thermal index, 32

Thermal pollution: definition, 3; technological controls, 7–8; problems, 243–248; future outlook, 263–271; misnomer, 281; prevention and costs, 294; basic principles, 303. *See also* Thermal discharges

Thermal power industry: Percentage of water withdrawn for cooling, 5; and regulation of plant construction, 11

Thermal races, 26

Thermal shock, 323

Thermal survey equipment, 209

Thermal threshold, 32

Thermal wedge, 157, 158

Thermionic power generation, 7

Thermistors, 198, 220

Thermocline, 167, 170

Thermometers: for recording river temperature, 114

Thermonuclear fusion: for central-station generation of electricity, 318

Tidenham Nuclear Power Plant (United Kingdom), 184

TLm. *See* Median tolerance limit (TLm)

Toxicity: result of raised temperatures, 31; of pollutants, 41. *See also* Poisons

Toxic pollution, 28

Tracers: and coefficients of turbulent dispersion, 203

Transmission lines: proximity to cooling towers, 255

Trash racks: to protect water screens, 237

Treatment processes, 333

Tropical worms: *Branchiura sowerbyi,*

15; in heated water in temperate countries, 46

Trout: hatching prevented by too-warm water, 4; yearling, thermal resistance, 40. *See also* Brook trout; Lake trout; Rainbow trout

Troy test flume, 235

Turbine-blade cooling: high-temperature, 7

Turkey Point Coastal Power Plant (Florida), 245, 246, 247

TVA. *See* Tennessee Valley Authority Project

Twenty-four-hour LD50, 32

Twenty-four-hour LT50, 36

U.S. Geological Survey (1954,) 135, 137, 138, 308

Udall, Stewart: and water-quality criteria, 10, 86, 103   7

Ultimate lethal temperature, 32

Ultimate upper lethal temperature, 32

United Kingdom: modeling of heated-water discharges, 177–212

Upper lethal temperature, 32

Upper median lethal temperature, 36

Utah Power and Light Company, 249, 264

Vaires-sur-Marne Power Station (France), 112–113, 118, 120, 121

Valmont, Lake, 309

Vapor. *See* Fog

Vistula River, 48

Vitamins: in seasonal phytoplankton successions, 28

Wallingford Hydraulics Research Station: thermal surveys, 182, 189; instrumentation and automated control systems, 205–210

Washington, state of: warm-water cultivation of oysters, 8

Waste dilution: of power-plant discharges, 124; jet-mixing by a diffuser, 168; heat, 246; possibilities considered, 319. *See also* Thermal discharges

Waste heat: as desalination aid, 9

Waste water: ecological effects, 176

Water: as life medium, 16

Water chemistry: and increased temperatures, 50–52

Water-level transmitters, 207

Water quality: temperature standards, 72–77; standard-setting process, 78; implementation of standards, 99; mentioned, 9–11, 72–109

Water Quality Act of 1965, 9, 72

Water screens. *See* Screens, water

Water temperature: ecological significance, 15–57; influences distribution and seasonal incidence of organisms, 23–26; seasonal fluctuations, 25; in aquatic ecosystems, 28, 322; effect on threshold concentration of poisons, 41; acclimation, 42, 65; lethal, 62; measurements, 80; differentials, 81; daily cycle, 128; diurnal range, 133; dry-bulb and wet-bulb, 273; economic significance, 282; problem of determining requirements, 324

Water temperature equilibrium. *See* Equilibrium temperature

Water temperature tolerance: range, 23; preferenda, 42; selection, 42; prediction, research needed, 331

Water temperature standards: determination of criteria for, 63–68; development, 73–75; Department of Interior reaction to, 75–76; monitoring, 76; administration, 76–77; philosophy, 78–79

Water treatment costs, 321

Wet-bulb temperature, 273

Wheeler Dam, 64

Wheeler Reservoir, 64

Widows Creek Steam Plant (Alabama), 80

Windage losses: from cooling devices, 284

Wind velocity: effect on cooling-tower performance, 274

Windermere, Lake, 25

Wood deterioration: in cooling towers, 259–260

Yellow perch: temperature selection, 42

Zinc: threshold of toxic concentration, 41

Zooplankton: seasonal successions, 26; and water temperature, 61–63